# *Essential Biology*

# Reproduction
# &
# Development

3rd edition

<u>STERLING</u>
Education

3   2   1

ISBN-13: 979-8-8855718-1-4

Sterling Education materials are available at quantity discounts.

Contact info@sterling–prep.com

Sterling Education
6 Liberty Square #11
Boston, MA 02109



Published by Sterling Education

 Printed in the U.S.A.

# STERLING
## Education

From the foundations of a living cell to the complex mechanisms of gene expression, *Essential Biology Self-Teaching Guides* are a comprehensive compendium of clearly explained texts to learn and master multifaceted biology topics.

These guides provide a detailed review of essential biological processes of living systems. Develop a better understanding of cell and molecular biology, mechanisms of cell metabolism, plants and photosynthesis, evolution and natural selection, ecology and population biology. Learn the principles of genetics, microbiology, classification and diversity, as well as structure and function of anatomical systems. Reinforce your learning by working through the practice questions and detailed explanations.

Created by highly qualified biology instructors, researchers, and education specialists, these books empower readers by helping them increase their understanding of biology.

We sincerely hope that these guides are valuable for your learning.

230801akp

## *Essential Biology Self-Teaching Guides*

Eukaryotic Cell & Cellular Metabolism

Molecular Biology & Genetics

Nervous & Endocrine Systems

Circulatory, Respiratory & Immune Systems

Digestive & Excretory Systems

Muscle, Skeletal & Integumentary Systems

Reproduction & Development

Microbiology

Plants & Photosynthesis

Evolution, Classification & Diversity

Ecology & Population Biology

**Visit our Amazon store**

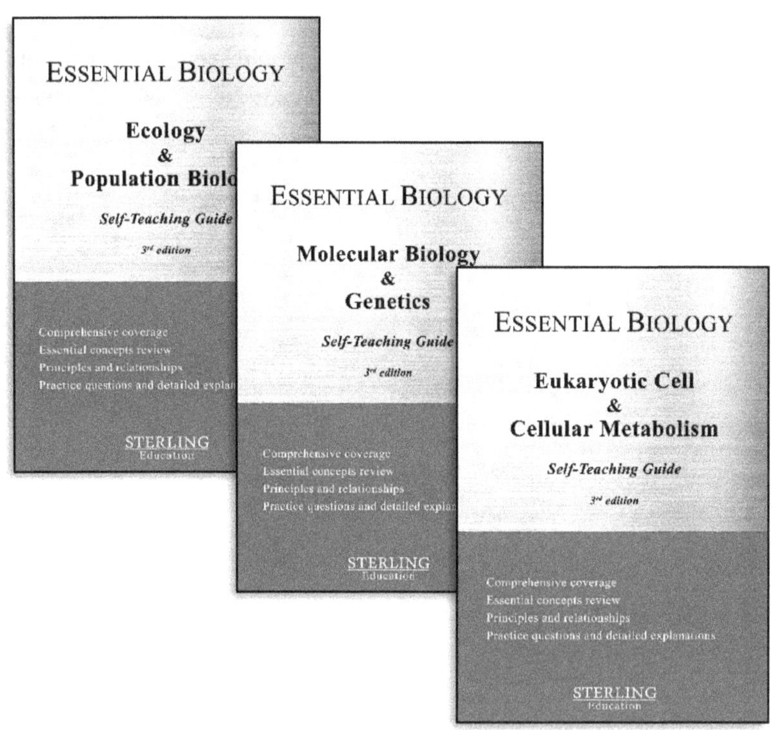

Kinematics and Dynamics

Equilibrium and Momentum

Force, Motion, Gravitation

Work and Energy

Fluids and Solids

Waves and Periodic Motion

Light and Optics

Sound

Electrostatics and Electromagnetism

Electric Circuits

Heat and Thermodynamics

Atomic and Nuclear Structure

**Visit our Amazon store**

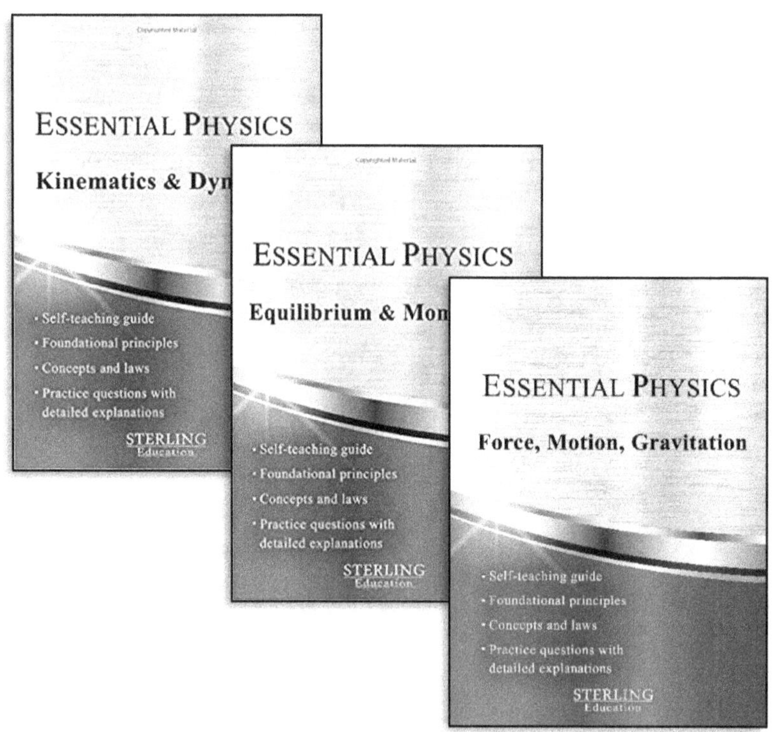

Chemistry

Physics

Cell and Molecular Biology

Organismal Biology

American History

American Law

American Government and Politics

Comparative Government and Politics

World History

European History

Psychology

Environmental Science

Human Geography

**Visit our Amazon store**

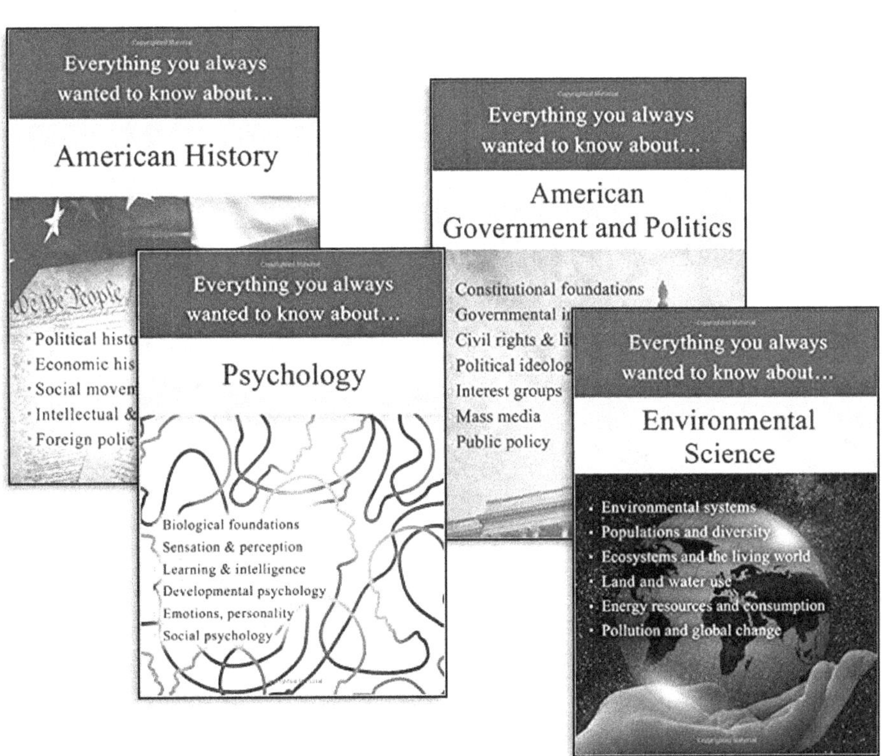

# Table of Contents

# Table of Contents (*continued*)

# Table of Contents (*continued*)

## **Table of Contents** (*continued*)

# Table of Contents (*continued*)

**Review: Development** (*continued*)

# Table of Contents (*continued*)

# Table of Contents (*continued*)

**REVIEW: Development** (*continued*)

## Table of Contents (*continued*)

# REVIEW

---

## Reproduction

## Female Reproductive Structures and Functions

### Genitalia of the reproductive system

*Genitalia* is the external genital organs of the reproductive system.

*Female reproductive system* includes the ovary, oviduct, uterus, and vagina.

The major difference between the male and female reproductive structures is that male structures are mostly external for the delivery of sperm.

Female structures are mainly internal to nurture a growing fetus.

### Female genitalia

*vulva* is the collective external genitalia of females. The urethra opens into the vulva.

*Vulva* (pudendum or external genitalia) includes the *clitoris, mons pubis, labia majora,* and *labia minora.* The labia are on each side of the vaginal and urethral openings.

*Clitoris* is a short shaft of erectile tissue capped by a pea-shaped gland at the front juncture of the labia minora. This structure is homologous to the male penis.

Additionally, the vulva contains the vaginal orifice, *greater* and *lesser vestibular glands, paraurethral glands,* and *vestibular bulbs* (erectile tissues).

### Vagina

*Vagina* is a tubular organ at a 45° angle with the small of the back.

Vagina is connected to uterus by the cervix, a cylinder-shaped neck of tissue 1 inch across.

*Cervix* is cartilage covered by smooth, moist tissue.

Vagina *tilts posteriorly* between the urethra and rectum.

It has no glands but is moistened by the transudation of serous fluid through the vaginal wall and mucus from glands in the cervical canal.

*Stratified squamous epithelium* with antigen-presenting *dendritic cells* line the vagina.

Vagina's mucosal lining lies in folds, extending as necessary in childbirth.

Vagina receives the penis during copulation.

*Copulation* is a sexual union that facilitates the reception of sperm by a female.

During birthing, the vagina is where the fetus passes out of the body (i.e., the birth canal).

## Uterus

*Uterus* has upper *fundus,* middle *corpus* (body), and lower *cervix* (neck), meeting the vagina.

A narrow *cervical canal* connects the uterine lumen with the vaginal lumen.

*Cervical glands* in the canal secrete mucus, preventing vaginal microbes from spreading into the uterus.

*Uterus* is a hollow, thick-walled muscular organ superior to the urinary bladder; its size and shape are comparable to an inverted pear, where the fertilized ovum develops until birth.

*Uterine wall* has three layers:

> an outer serosa *perimetrium,*
>
> thick muscular *myometrium,* and
>
> an inner mucosa *endometrium.*

Endometrium contains tubular glands divided into two layers –

> thick superficial *stratum functionalis* (shed during menstrual periods) and
>
> thinner basal *stratum basalis* (retained from cycle to cycle).

The uterus is supported by a pair of lateral wing-like *broad ligaments* and cordlike *cardinal, uterosacral,* and *round ligaments.* It receives blood from a pair of *uterine arteries.*

## Fallopian tubes

*Fallopian tubes* (or *oviducts*) are two tubes that branch from the uterus and provide a passage to the uterus for the ovum released by the ovary.

Fallopian tubes, uterine tubes, and salpinges (singular salpinx) are the expected fertilization sites. They are lined with ciliated epithelia.

After fertilization, the embryo is slowly moved by ciliary movement toward the uterus.

The flared distal end of the Fallopian tube, near the ovary, is the *infundibulum* and has feathery projections of *fimbriae* to receive the ovulated egg.

Its long midportion is the *ampulla,* and a short, constricted zone near the uterus is the *isthmus. Mesosalpinx* is a ligamentous sheet that supports the tube.

**Female reproductive structures**

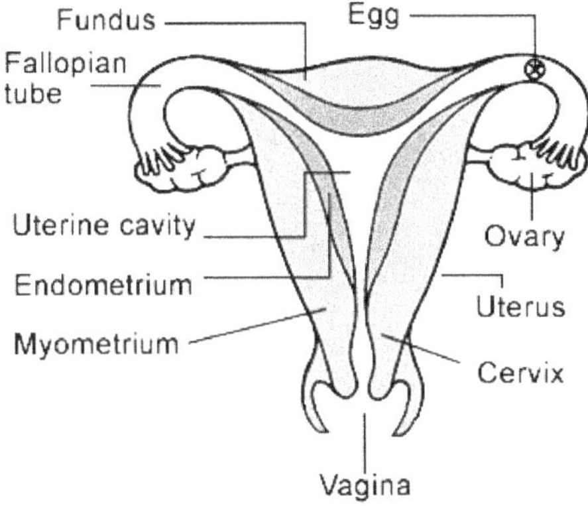



*Female reproductive system with associated structures*

**Gonad anatomy and functions**

Female gonads are *ovaries* with immature eggs that mature one (or more) at a time (monthly).

*Ovary* is where ova (or eggs) are produced (ovaries produce a secondary oocyte each month) along with female sex hormones, *estrogen*, and *progesterone*, during the ovarian cycle.

Ovaries are in the *abdominal cavity*.

Females typically have two ovaries.

*Oviduct* (Fallopian or uterine tube) is a tube for egg movement from the ovary to the uterus.

**Ovarian anatomy**

Each ovary is associated with an *oviduct* and has the following:

> a central *medulla*,

> a surface *cortex,* and

> *tunica albuginea* as an outer fibrous capsule.

*Fimbriae* are finger-like projections that sweep over the ovaries and waft the egg into the Fallopian tubes when released from the ovary.

Generally, the ovaries alternate in producing one oocyte every month.

## Ligaments

Ovary is supported by:

medial *ovarian ligament,*

lateral *suspensory ligament,* and

anterior *mesovarium.*

## Ovarian blood supply

Ovary receives blood from a branch of *uterine artery* medially and *ovarian artery* laterally.

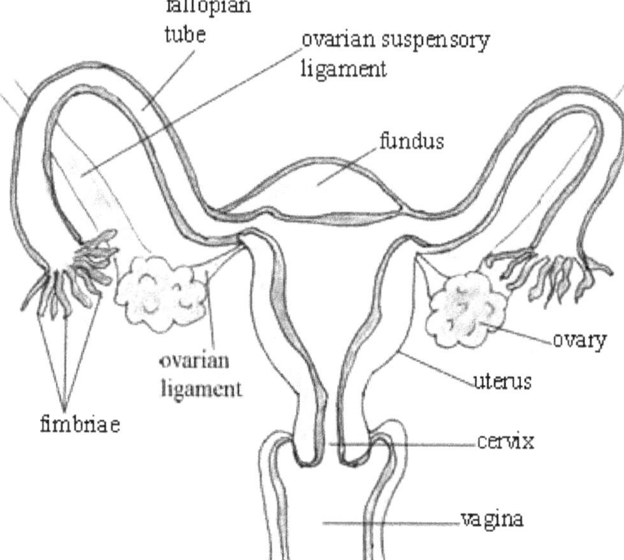


*Female ovary with fallopian tube and ovarian ligament*

## Male Reproductive Structures and Functions

### Male genitalia

*Genitalia* is the external genital organs of the reproductive system. External genitalia of men consists of the *penis, male urethra*, and *scrotum.*

*Male reproductive system* consists of the testes, epididymis, vas deferens, prostate gland, bulbourethral glands, and penis.

Males have one opening for urine and sperm, while females have separate openings for urine, menstruation, and sexual intercourse.

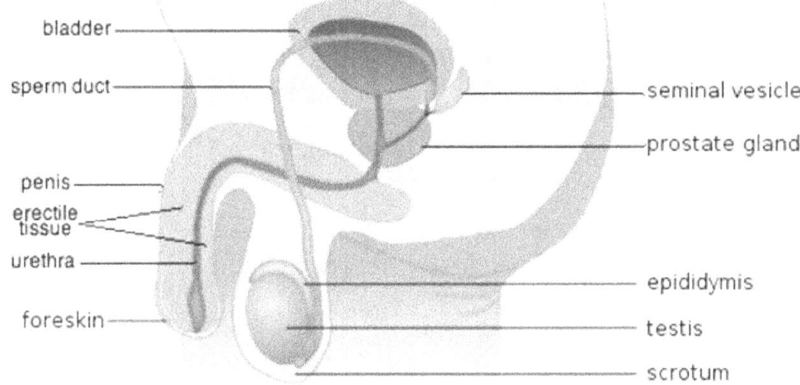


*Male reproductive system with anatomy and glands*

### Penis

*Penis* is a cylindrical copulatory organ that introduces semen (a fluid containing spermatozoa) and secretions into the female vagina.

The penis is divided into an internal *root* and an external *shaft* and *glans*. It is covered with loose skin that extends over the glans as the *prepuce* (foreskin).

Internally, the penile shaft consists mainly of three spongy long *erectile tissues*:

pair of dorsal *corpora cavernosa* (engorge with blood and effect of erection) and a

single ventral *corpus spongiosum* (contains the urethra).

All three tissues have *lacunae* (or *blood sinuses*) separated by *trabeculae* composed of connective tissue and *trabecular muscle* (or *smooth muscle*).

At the proximal end of the penis, the *corpus spongiosum* dilates into a *bulb* that receives the urethra and ducts of the *bulbourethral glands*. *Corpora cavernosa* diverges into a pair of *crura,* anchoring the penis to the pubic arch and perineal membrane.

## Blood supply

A pair of internal *pudendal arteries* supply the penis.

*Pudendal arteries* branch into a *dorsal artery*, travelling dorsally under the skin of the penis.

*Deep artery* travels through the corpus cavernosum and supplies blood to the lacunae.

*Dorsal arteries* supply most of the blood when the penis is flaccid, and the deep arteries supply blood during an erection.

## Scrotum

*Scrotum* is part of the external male genitalia behind and underneath the penis. It is a pouch of skin surrounding and protecting the testicles.

The scrotum contains the testes and the *spermatic cord*, a bundle of *connective tissue*, *testicular blood vessels*, and *ductus deferens* (i.e., *sperm duct*).

*Spermatic cord* passes up the back of the scrotum and through the external inguinal ring into the inguinal canal.

## Spermatic ducts

*Spermatic ducts* carry sperm from the *testes* to the *urethra*.

Spermatic ducts include:

> *efferent ductules* (leaving the testes);
>
> *duct of epididymis* (highly coiled structure adhering to the posterior side of testes);
>
> muscular *ductus deferens* (travels through the spermatic cord and inguinal canal into the pelvic cavity); and a
>
> short *ejaculatory duct* (carries sperm and seminal vesicle secretions towards the last 2 cm to the urethra).

*Urethra* completes the path of the sperm to the outside of the body.

Urethra is the duct by which *urine* is conveyed out of the body from the *bladder* and by which *male vertebrates convey semen.*

## Gonads

Male gonads are the *testes* that produce sperm and testosterone.

Paired testes are suspended in the scrotal sacs of the scrotum.

Shortly before birth, the fetal testes descend through the inguinal canal into the scrotum.

Low temperature in the scrotum is vital to normal sperm production.

If testes do not descend, surgery or hormonal therapy is required; otherwise, sterility results.

Testis consists of *seminiferous tubules* to produce sperm and *interstitial cells* (Leydig cells) to produce testosterone.

Seminiferous tubule epithelium consists of *germ cells* and *sustentacular cells.*

*Germ cells* develop into sperm.

*Sustentacular cells* support and nourish germ cells by forming a *blood-testis barrier* between them and the nearest blood supply.

Like the ovary, the testis has a fibrous capsule of the *tunica albuginea.*

*Fibrous septa* extend from the *tunica* and divide the interior of the testis into 250–300 compartments of *lobules.*

Lobules contain one to three sperm-producing seminiferous tubules.

Testosterone-secreting interstitial cells lie in clusters in between the tubules.

## Gonad anatomy and function

A long, slender *testicular artery* supplies each testis and is drained by pampiniform plexus veins, which converge to form the *testicular vein.*

Testis is supplied with *testicular nerves* and lymphatic vessels.

*Epididymis* is a coiled tube attached to each testicle for *maturation and storage of sperm.*

*Vas deferens* and *epididymis* store sperm until ejaculation.

Sperm is *non-motile* while stored in the *vas deferens* and *epididymis*. During the passage through the epididymis, they are concentrated by fluid absorption.

When a male is sexually aroused, the sperm enters the urethra, extending through the penis.

Sperm travels the *vas deferens* into the *ejaculatory duct,* which leads to the *urethra* and *penis.*

*Urethra* transports urine from the *bladder* during urination.

**Blood supply and innervation**

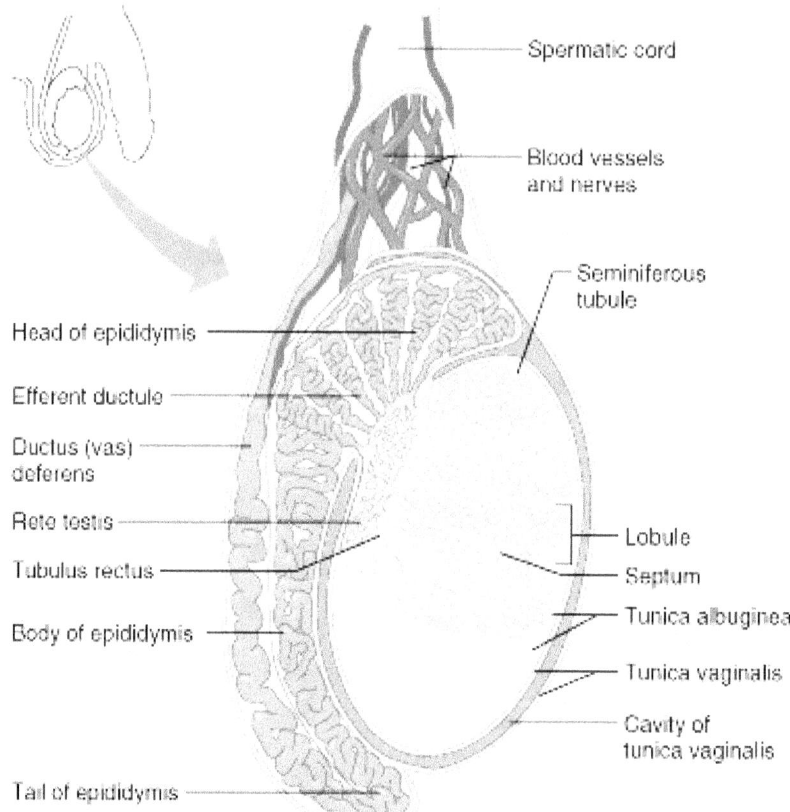


*Male reproductive system with blood vessels, ducts and tissue shown*

**Accessory glands**

Males have *three sets* of accessory glands:

a pair of *seminal vesicles* posterior to the urinary bladder,

a pair of small *bulbourethral glands* secreting into the proximal end of penile urethra.

a single *prostate gland* inferior to the bladder (enclosing the prostatic urethra) and

*Seminal vesicles* lie at the urinary bladder base, which contains two glands that join the *vas deferens* (pl. *vasa deferentia*) to form an ejaculatory duct that enters the urethra.

A thick fluid containing nutrients, including mucus (liquid for the sperm), fructose (for ATP), and prostaglandins, is secreted into the ejaculatory duct.

*Bulbourethral glands* are below the prostate gland and on either side of the urethra; they release mucous secretions that provide lubrication.

Bulbourethral glands produce a small amount of clear, slippery fluid that lubricates the urethra and neutralizes its pH.

*Prostate gland* is below the urinary bladder and surrounds the upper portion of the urethra.

Prostate secretes a milky, slightly alkaline solution that promotes sperm motility and viability. This fluid neutralizes urine acidity in the urethra and neutralizes vaginal acidity.

*Prostaglandin secretions* neutralize seminal fluid, acidic from the metabolic waste of sperm.

## Semen

*Semen* (seminal fluid) is a thick, whitish fluid containing about:

> 10% sperm,
>
> 30% glandular secretions from the prostate vesicles, and
>
> 60% secretions from the seminal vesicles and bulbourethral glands.

Semen contains about 50–120 million sperm/mL and *seminogelin* (seminal vesicle protein), a serine protease (prostate-specific antigen), fructose, prostaglandins, and other substances.

The seminal vesicles and prostate secrete most of the semen.

## Accessory muscles

Sperm cannot develop at the core body temperature of 37 °C.

Testis is about 2 °C cooler than average body temperature by three structures in the scrotum:

> *cremaster muscle* of the spermatic cord,
>
> *dartos muscle* in the scrotal wall, and the
>
> *pampiniform plexus* of veins in the spermatic cord.

*Cremaster muscle* relaxes when warm and contracts when cool, lowering or raising the scrotum and testes.

*Dartos muscle* contracts and tautens the scrotum when it is cool.

*Pampiniform plexus* is a countercurrent heat exchanger cooling blood on its way to the testes.

## Sperm pathway

Mnemonic for the path of sperm:

"Seven Up" (<u>S</u>eminiferous tubules, <u>E</u>pididymis, <u>V</u>as deferens, <u>E</u>jaculatory duct, <u>n</u>othing, <u>U</u>rethra, <u>P</u>enis).

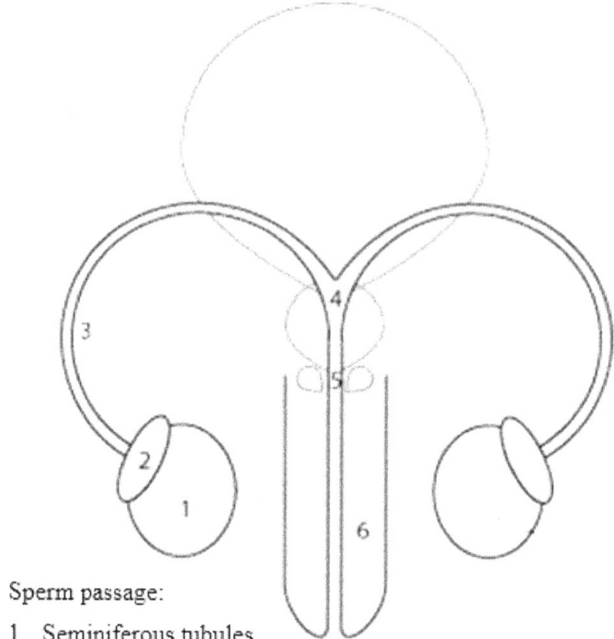


Sperm passage:

1. Seminiferous tubules
2. Epididymis
3. Vas deferens
4. Ejaculatory duct
5. Urethra
6. Penis

*Path of sperm from the seminiferous tubules and through the penis as "Seven Up"*

 

## Gametogenesis

### Gonads produce gametes

Gonads are *primary sex organs* specialized to produce *gametes* (haploid egg and sperm).

Two types of gonads:

> *testes*, which produce spermatozoa (sperm), and

> *ovaries*, which produce ova (singular ovum or egg).

### Gamete formation

*Gametogenesis* is the meiotic cell division that produces eggs (oogenesis) and sperm (spermatogenesis).

*Meiosis* is a nuclear division that is broken into two broad stages (meiosis I and meiosis II), which reduces the chromosome number from *diploid* (2N) to *haploid* (N).

The haploid number is half of the diploid number of chromosomes.

Meiosis of one diploid cell produces four haploid cells.

These daughter cells, containing half the genetic information, are *gametes.*

Before fertilization, the building blocks to create that cell must be available so that the initial *fertilized cell* (i.e., zygote) can develop into an embryo, fetus, and adult.

### Gamete development

Developing gametes (egg and sperm) are *germ cells.*

The first stage in their development is the proliferation of primordial germ cells by mitosis (replication via cell division); *daughter cells* receive a *complete set* of chromosomes identical to the original cell.

In females, germ cells' mitosis activity occurs during *embryonic development*, while in males, it begins at *puberty and continues throughout life.*

For years, females were believed to contain all their egg cells at birth. However, recent findings of mitotic activity in female gonads and the idea that females do not produce new eggs during their lifetime are actively researched.

**Meiosis**

The next stage of development is meiosis; each daughter cell receives half the chromosomes of the original cell.

During meiosis, chromosomes stay in their homologous pairs.

For example, instead of 46 individual human chromosomes, there are 23 chromosome pairs.

Gamete production via meiosis occurs in sexually reproducing eukaryotes, including animals, plants, and fungi.

Sexual reproduction forms gametes.

Human males produce tiny, motile sperm in the testes, and human females produce a large, immobile, nutrient-laden egg or ovum in the ovarian follicles.

Gametes fuse to form a *zygote* with the full or diploid (2N) number of chromosomes.

If gametes contained the same number of chromosomes as somatic (body) cells, the zygote would have twice the correct number of chromosomes.

**Diploid zygote**

When 1N gametes fuse, the chromosomes from both parents combine in the 2N zygote.

Each fertilized egg (i.e., zygote) cell has a pair of homologous chromosomes, one 1N homolog from the mother (*egg*) and one 1N homolog from the father (*sperm*).

Chromosomes align within the daughter cells in many combinations during meiosis. $(2^{23})^2$ or about 70 trillion combinations are possible without crossing over.

This allows for genetic variability in sexually reproducing organisms.

Crossing over occurs during meiosis I, where homologous chromosomes are paired (synapsis) and exchange some genetic material to form genetically unique (recombinant) chromosomes.

If crossing over occurs once, $(4^{23})^2$ or 70 trillion squared genetically unique zygotes are possible for one couple.

Crossing over is unique to prophase I of meiosis.

## Meiosis I & II

**Phases of meiosis I**

Meiosis I and meiosis II have four phases: prophase, metaphase, anaphase, and telophase.

Before meiosis I, DNA replication occurs in the S phase of interphase, and each chromosome has a pair of *sister chromatids.*

Sister chromatids are attached at the centromere (like mitosis for somatic cells).

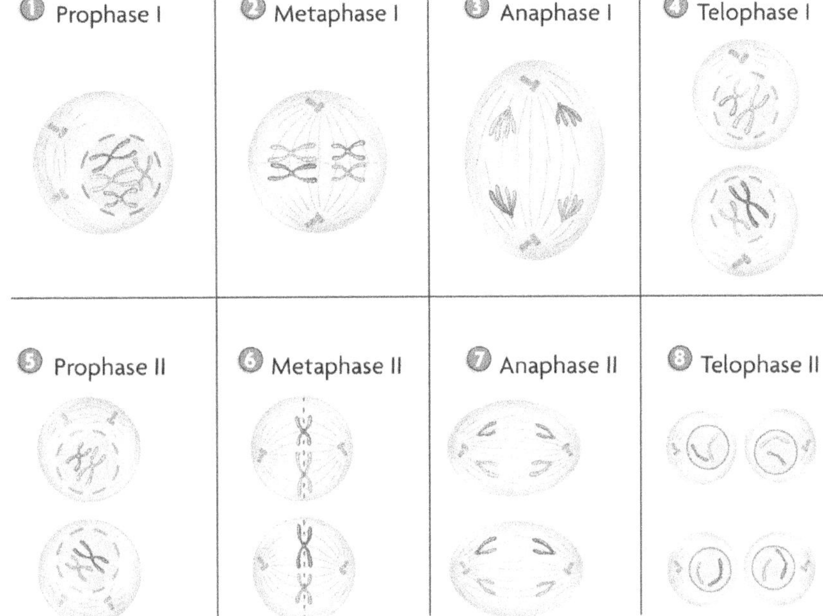


*Meiosis I is a reduction stage (2N → 1N), meiosis II separates sister chromatids at the centromere*

**Meiosis I: reduction phase from diploid to haploid**

During meiosis I, homologous chromosomes line up at the synapsis.

Two sets of paired chromosomes (1 from mother and 1 from father) are together as bivalents (tetrads) held by a chiasma complex.

**Prophase I**

1.  Nuclear division occurs: nucleolus disappears, nuclear envelope fragments, centrosomes migrate away from each other, and spindle fibers assemble.

2. Homologous chromosomes undergo synapsis, forming bivalents; crossing over may occur as sister chromatids exchange genetic material by recombination.

3. Chromatin condenses, and chromosomes are microscopically visible.

## Metaphase I

1. During prometaphase I, bivalents held by chiasmata move toward the metaphase plate at the equator of the cell.

2. The fully formed spindle aligns the bivalents at the metaphase plate.

3. Kinetochores are proteins associated with centromeres; they attach to *kinetochore spindle fibers* anchored to centrioles at each pole of the cell.

4. Homologous chromosomes independently align at the metaphase plate.

5. Maternal and paternal homologs may be oriented toward either pole.

## Anaphase I

1. The homologs separate and move toward opposite poles.

2. Each chromosome is attached with a centromere and two sister chromatids (replicated previously during the S phase).

## Telophase I

1. In animals, this occurs at the end of meiosis I.

2. The nuclear envelope reforms and nucleoli reappears.

3. This phase may or may not be accompanied by cytokinesis for portioning the two nuclei with separate plasma membranes (i.e., two daughter cells).

## Interkinesis

1. Between meiosis I and meiosis II is like interphase in mitotic divisions.

2. However, no DNA replication (as in S phase of mitosis) occurs; the chromosomes are 1N (each chromosome has a sister chromatid).

**Meiosis II: 1N → 1N**

Before meiosis II, DNA does *not* replicate, and centromere still attaches to sister chromatids.

During meiosis II, the centromeres split, and the sister chromatids separate.

Chromosomes in the four daughter cells contain one chromatid.

Counting the number of centromeres verifies the number of chromosomes.

**Fertilization restores diploid number**

Fertilization restores the diploid number (2N) in the zygote and somatic cells originating from a zygote.

1. During metaphase II, haploid chromosomes (with sister chromatids) align at the metaphase plate.

2. During anaphase II, sister chromatids separate at the centromeres, and the two daughter chromosomes move toward the poles.

3. Due to crossing over in prophase I, each gamete contains chromosomes with gene combinations, unlike either parent.

4. At the end of telophase II and cytokinesis, there are four haploid cells (1 sperm for males and 1 egg, and 3 polar bodies for females).

5. In animals, the haploid cells mature and develop into gametes, which may fuse into a zygote (2N) from a 1N sperm and 1N egg.

6. In plants, the daughter cells become spores and divide to produce a haploid adult generation.

7. In some fungi and algae, a zygote results from gamete fusion and immediately undergoes meiosis; therefore, the adult is haploid.

**Unique haploid cells**

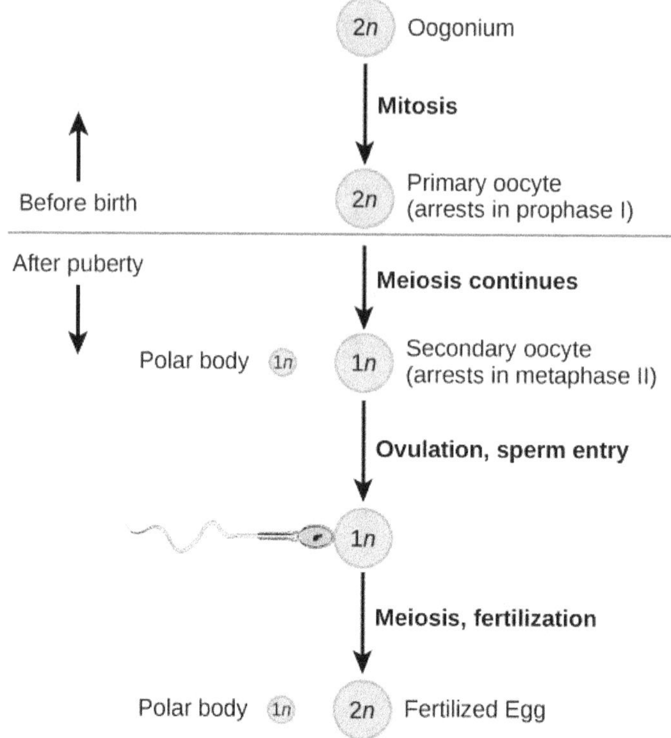



*Meiosis generates 4 unique haploid cells, while mitosis generates two identical diploid cells*

**Haploid gametes fuse during fertilization**

DNA is replicated once before mitosis and meiosis I and II;

in mitosis, there is *one nuclear division,*

in meiosis, there are *two nuclei divisions* between syntheses of DNA.

In humans, meiosis occurs in reproductive organs to produce gametes, while mitosis occurs in somatic cells (not germline cells) for growth and repair.

**Meiosis *vs*. mitosis**

- During **prophase I** of meiosis, homologous chromosomes pair for crossing over, increasing genetic variation.

  Crossing over does *not* occur during mitosis.

- During **metaphase I** of meiosis, *homologous chromosomes* align at the metaphase plate.

  In mitosis, individual chromosomes align.

- During **anaphase I** of meiosis, homologous chromosomes, with centromeres intact, separate and move to opposite poles.

  In mitosis, *sister chromatids separate* and move to opposite poles.

- During **meiosis II** of meiosis, the stages are the same as mitosis.

- Nuclei contain the *haploid* (1N) number of chromosomes in meiosis.

  Cell contains the *diploid* (2N) number of chromosomes in mitosis.

- Meiosis produces *four* genetically unique *haploid daughter cells* (4 sperm or 1 egg and 3 polar bodies).

  Mitosis produces *two* genetically identical *diploid daughter cells*.

*Notes for active learning*

## Oogenesis

### Ovarian cycle

Oogenesis produces a *single ovum* from a *single primary oocyte*.

Unlike spermatogenesis, the *ovarian cycle* occurs monthly and usually produces one gamete (egg) per month.

Each egg develops in its bubble-like *follicle*, located primarily in the cortex.

Each month, about 20 to 25 *primordial follicles* resume their development.

The single layer of squamous follicular cells around the oocyte thickens into cuboidal cells.

The follicle is a *primary follicle.*

As the egg enlarges, the follicular cells multiply and aggregate into multiple strata; the follicle is a *secondary follicle*, and the follicular cells are *granulosa cells.*

### Oogenesis *vs.* spermatogenesis

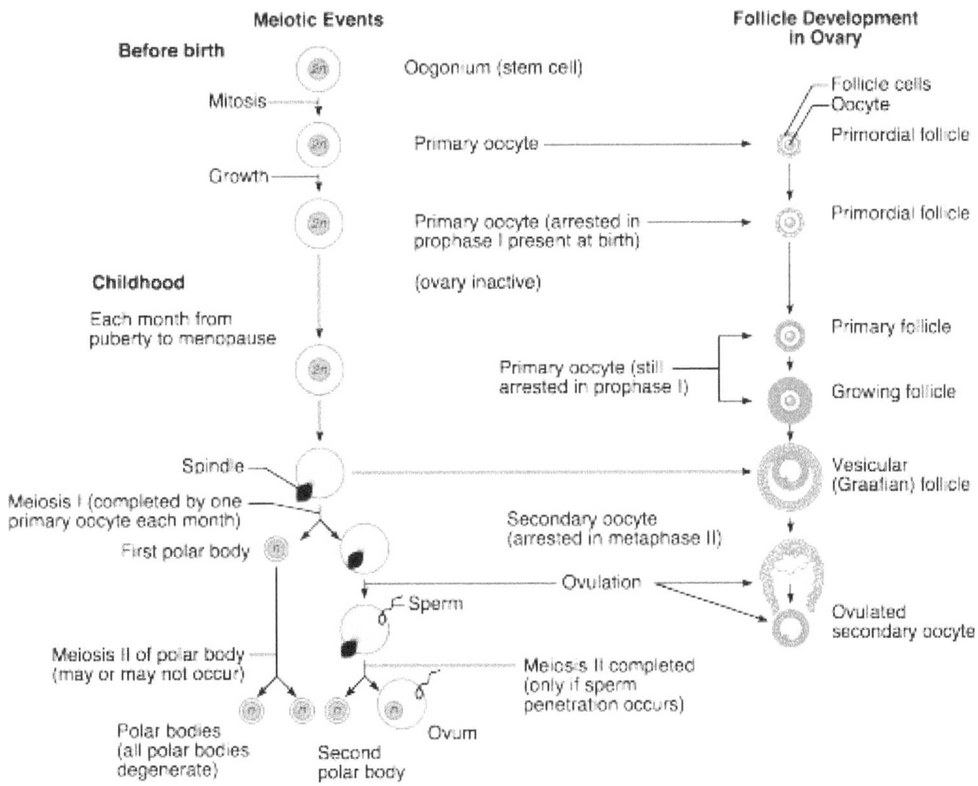

## Oocytes

*Oogonia* are primitive germ cells that undergo mitosis and develop into *primary oocytes* (with 46 chromosomes), the initial cells in oogenesis.

Primary oocytes remain in meiotic arrest (i.e., begin first meiotic development but not completed).

The germ cells in a female are believed to be at this developmental stage at birth.

However, this is an area currently under investigation by researchers.

Primary oocytes are in the ovaries of the female reproductive system.

Oogenesis occurs monthly, beginning at puberty and ending at menopause.

At puberty, primary oocytes are destined for ovulation and complete meiosis, and each daughter cell receives 23 chromosomes.

Although some primary oocytes undergo *atresia* (immature and degraded) during childhood, there are about 300,000 to 400,000 oocytes at puberty.

When the primary oocyte divides, one of the two daughter cells, the *secondary oocyte*, retains most cytoplasm.

## Polar bodies

*First polar body* is the nonfunctional daughter cell and (often) does not proceed to meiosis II.

The secondary oocyte proceeds to metaphase II of meiosis and suspends.

Meiosis II resumes if fertilization occurs.

Completing meiosis II allows the secondary oocyte to become a fertilized egg (2N zygote when fused with the sperm).

Meiotic division produces two *second polar bodies* that disintegrate because they receive insufficient cytoplasm.

The body absorbs second polar bodies and retains most of the egg's cytoplasm.

The cytoplasm serves as a source of nutrients for the developing embryo.

This is different from spermatogenesis, in which four functional, mature sperm are produced from a single spermatogonium.

## Ovum development

An oocyte is a *developing egg.*

Oocytes undergo two separate *meiotic cell divisions* before becoming a mature *ovum.*

Oocytes differentiate into a mature egg (or ovum) by meiosis when germ cells complete two final, highly specialized divisions.

A single ovum is produced per month.

## Ovulation

Typically, one tertiary follicle becomes a fully mature *vesicular* (Graafian) or *tertiary follicle* destined to ovulate.

Ovulation occurs around day 14 of a typical cycle.

Follicle swells and bursts, releasing the egg and *cumulus oophorus* (a mass of follicular cells surrounding the ovum in the vesicular ovarian follicle) into the opening of the uterine tube.

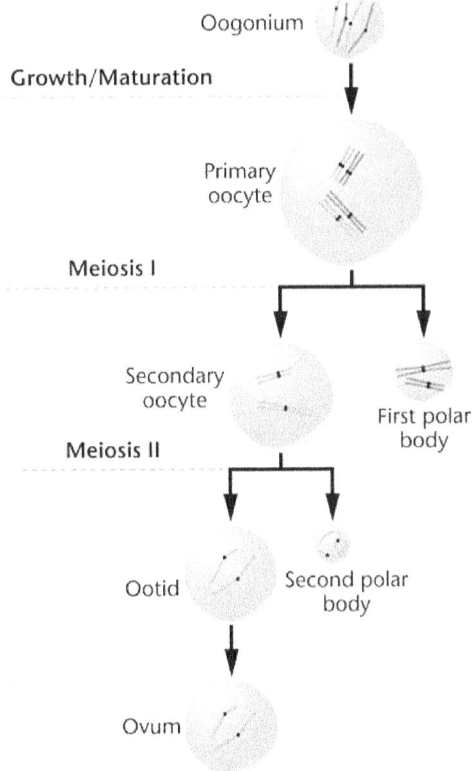


*Female oogenesis with 1 ovum and (up to) 3 polar bodies*

*Notes for active learning*

# Spermatogenesis

## Sperm formation

In human males, meiosis is part of *spermatogenesis*, producing sperm in the testes.

In human females, meiosis is in *oogenesis*, the production of egg cells occurring in ovaries.

*Spermatogenesis* occurs in the *seminiferous tubules* in the testes and produces *sperm* from *primary spermatocytes*.

## Primary and secondary spermatocytes

*Spermatogonia* are undifferentiated germ cells that divide by mitosis and differentiate into *primary spermatocytes*.

Primary spermatocytes grow and undergo the *first meiotic division* to form *secondary spermatocytes*.

## Spermatid formation

In secondary spermatocyte stage, cells undergo a *second meiotic division* to form *spermatids*.

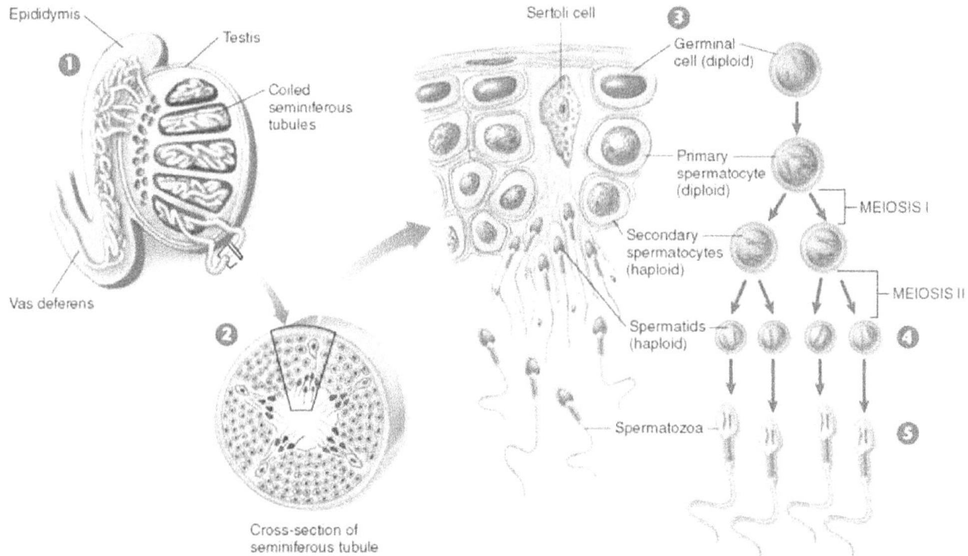


*Testis and spermatogenesis. Mitosis (2N) is followed by meiosis I (1N) and meiosis II (1N)*

**Spermatogonium**

Spermatogonium replicates all its chromosomes during interphase.

Spermatogonium has 46 primary spermatocyte chromosomes, and each chromosome is two sister chromatids joined by a centromere.

Cells undergo prophase I, where homologous chromosomes line up on spindles at the equator.

Cells undergo anaphase I (reduction phase of meiosis), but the centromeres do not divide; the *homologous chromosome pairs* separate.

During telophase I, two cells are formed.

Each cell is a secondary spermatocyte that contains 23 chromosomes (i.e., haploid cells), and each chromosome has *two chromatids* joined by a centromere.

**Meiosis II**

Haploid cells undergo meiosis II:

> prophase II, metaphase II, anaphase II, and telophase II.

This second set of divisions *does* resemble mitosis.

Chromosomes condense (but do not pair up) during prophase II.

Chromosomes align by spindle fibers during metaphase II, and the centromeres divide during anaphase II. The cells finish dividing during telophase II.

At end of meiosis II, there are *four haploid cells* with 23 chromosomes.

*Spermatids*, formed from the division of the secondary spermatocytes, develop into mature *spermatozoa* (or *sperm*).

**Sertoli cells**

*Sertoli cells* are stimulated by follicle-stimulating hormone (FSH) in the seminiferous tubules and surround and nourish spermatids during differentiation.

*Spermatids* complete maturation (e.g., a gain of motility) in the *epididymis*.

*Sertoli cells* secrete peptide hormones:

> *inhibin* (acts on the pituitary gland to inhibit FSH release) and

> *androgen-binding protein* (binds testosterone and is an intermediary between germ cells and hormones).

Sertoli cells divide the tubules into compartments with different environments where stages of spermatogenesis continue.

                                          

**Leydig cells and epididymides**

*Leydig cells* between tubules produce testosterone in presence of luteinizing hormone (LH).

Sperm produced in the testes mature within the *epididymides.*

*Epididymides* are tightly coiled tubules outside of the testes.

Maturation time in the epididymis is required for sperm to develop the ability to swim.

Once sperm matures, they are propelled into the *vasa deferentia* by muscular contractions.

Sperm is stored in the epididymides and the vasa deferentia.

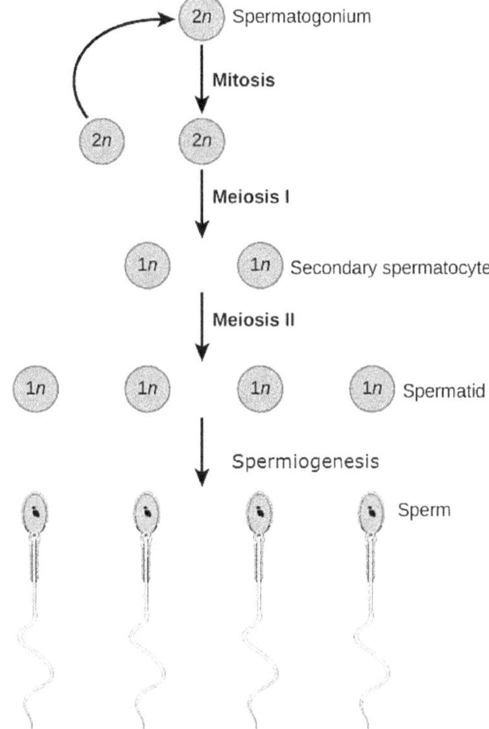


*Male spermatogenesis with haploid secondary spermatocytes (1N) and spermatids (1N)*

*Notes for active learning*

## Gamete Formation

### Female *vs.* male gametes

*Ovum* (or unfertilized egg cell) is the female gamete.

Unlike sperm, the egg is not capable of active movement.

The egg is much larger than the sperm (visible to the naked eye).

Human sperm is about 55 micrometers ($\mu$m) long (head is 5 $\mu$m, and flagellum is 50 $\mu$m).

A mature ovum is between 120-150 $\mu$m in diameter.

Therefore, the ratio of the length of the sperm to the diameter of the egg is about 1:3.

| Male | Female | Difference |
|---|---|---|
| Spermatogonium (2N) | Oogonium (2N) | Spermatogonium renews its population by mitosis throughout life. Oogonium stops renewing its population before birth |
| Primary spermatocyte (1 N) | Primary oocyte (1N) | Primary oocyte arrests at prophase I |
| Secondary spermatocyte (1N) | Secondary oocyte (1N) | Secondary oocyte arrests at metaphase II |
| Sperm (1N) | Ovum (1N) | Between the secondary spermatocyte and sperm is the spermatid |

## Egg morphology

Compared with the width of a sperm cell (~3 μm) to an egg's diameter, the ratio is about 1:50.

Eggs are non-motile and filled with cytoplasm.

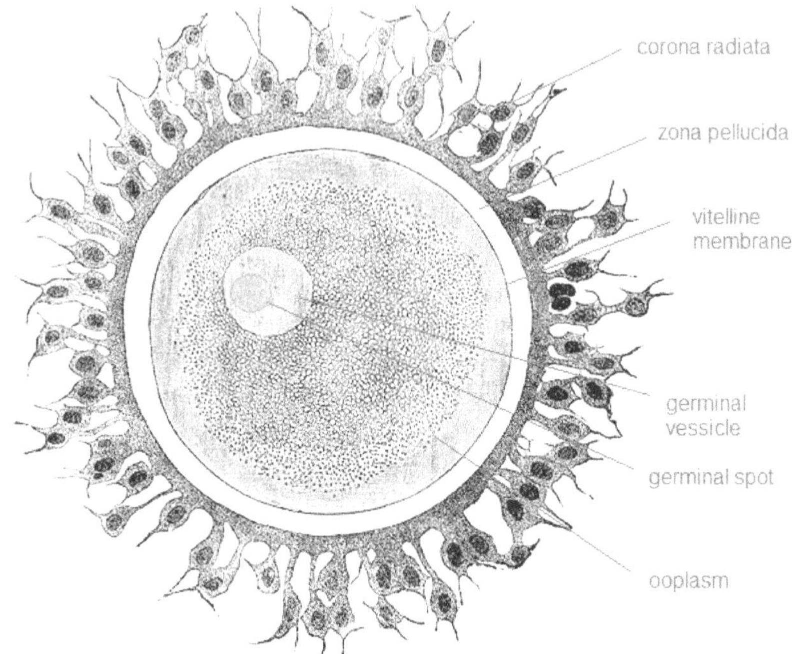


*Human ovum with the protective glycoprotein layer of zona pellucida
and inner vitelline membrane enclosing the ooplasm*

## Granulosa cell

*Granulosa cells* produce a glycoprotein gel layer as the *zona pellucida* around the egg; the connective tissue around the granulosa cells condenses into a tough fibrous *theca folliculi.*

Granulosa cells secrete *follicular fluid*, which forms small pools.

The follicle is *tertiary* with continued proliferation of granulosa and theca cells, increased thecal vascularization, and further oocyte enlargement.

Ovary functions as an endocrine organ when transitioning from secondary to tertiary follicles.

Fluid pools eventually coalesce to form a single cavity, the *antrum.*

The egg is held against one side of the antrum by a mound of cumulus oophorus cell*s*; the innermost layer of these cells is the *corona radiata.*

## Sperm morphology and mechanisms

Sperm is compact cells of haploid male DNA with flagella that provide motility.

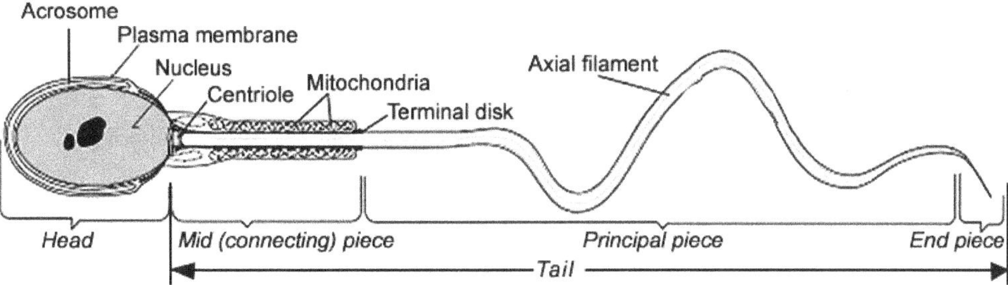


*Human spermatozoon with head, midpiece, and tail with mitochondria for motility*

## Spermatozoa

*Spermatozoa* (i.e., mature sperm) have the following.

*Sperm head* (haploid with 23 chromosomes) has a nucleus and DNA covered by an acrosome.

*Acrosome* is a cap-like covering over the anterior end of the nucleus that stores enzymes to help sperm penetrate the several layers of cells and thick membrane enclosing the egg.

*Middle piece* contains multiple mitochondria wrapped around microtubules of the flagellum in a $9 + 2$ microtubule array for energy.

*Tail* has microtubules as a flagellum for its whip-like movement, propelling sperm.

Ejaculate of a human male contains several hundred million sperm.

Fewer than 100 reach the vicinity of an egg, and usually, only one sperm enters an egg.

## Egg and sperm contributions during fertilization

*Sexual reproduction* produces offspring that combine genes from each parent.

It is the union of two gametes (i.e., *ovum* and *sperm*) to form a zygote (i.e., *fertilized egg*).

Gametes contain half the genetic information needed to produce the 2N progeny.

Sperm contributes chromosomal DNA, as the egg actively destroys *sperm mitochondria*.

Egg contributes chromosomal DNA and mitochondria, organelles, and biomolecules within the ooplasm within the large volume of the cytoplasm.

Cytoplasmic volume requirements drive the partitions of gametes into 1 large egg and 3 small, non-functional *polar bodies*.

## Cytoplasmic nutrients and biomolecules

Egg contains *cytoplasm nutrients* and *biomolecules* needed by a developing embryo.

A diploid zygote (haploid sperm and haploid egg) represents the first stage in developing a genetically unique organism.

Zygote contains essential components for development with genes on chromosomes.

Zygote genes are not activated to produce proteins until after several cleavage divisions.

During cleavage, the large zygote subdivides into cells of conventional size via mitosis.

After preliminary cleavage, blastomeres are the initial cells of an organism's development.

## Sexual Development

### Male and female sexual maturity

Ovaries produce *estradiol* and *progesterone.* Estradiol is the female hormone estrogen.

Testes produce *testosterone,* which is the main sex hormone in males. Testosterone is an androgen, which has masculinizing effects.

Androgens are not unique to males, and estrogens are not unique to females.

Many male and female reproductive system organs develop from the same embryonic organs.

Organs with the same embryonic precursor are *homologous.*

For example, the scrotum and labia majora are homologous because each develops from the labioscrotal folds.

*Secondary sex organs* are anatomical structures needed to produce offspring, such as the male glands, ducts, penis, female uterine tubes, uterus, and vagina.

*Secondary sex characteristics* are not essential to reproduction but help attract mates by indicating sexual maturity (e.g., breasts).

Secondary sex characteristics comprise the *external differences* between males and females.

### Reproductive organs mature during puberty

*Puberty* is when reproductive organs mature, and reproduction becomes possible.

Puberty is initiated by the secretion of *gonadotropin-releasing hormone.*

(GnRH) by the hypothalamus. GnRH stimulates the release of *follicle-stimulating hormone* (FSH) and *luteinizing hormone* (LH) by the anterior pituitary.

FSH and LH act upon gonads to stimulate the development of sperm and ova and the secretion of sex hormones.

Sex hormones exert negative feedback on the secretion of GnRH, FSH, and LH.

Visible changes at puberty are from hormones (e.g., testosterone, estrogens, growth hormone).

After puberty, the individual attains fertility.

Adolescence continues until full adult height is attained.

 

**Secondary male characteristics**

The earliest visible sign of male puberty is an enlargement of the testes and scrotum; the ejaculation of motile sperm marks the completion of puberty.

Testosterone has three prominent roles:

1) stimulating the development of prenatal genitalia,

2) stimulating the development of male secondary sexual characteristics, such as the growth of skeletal muscle (longer legs and broader shoulders), development of pubic hair, facial hair, and chest hair, and

3) maintaining sex drive during adulthood. It stimulates the secretion of oil and sweat glands (i.e., attributed to body odor and acne).

Testosterone prompts the *larynx* and *vocal cords* to enlarge, resulting in a deeper voice.

Testosterone triggers baldness if "baldness genes" are present and regulates testosterone synthesis by acting on Leydig cells to stimulate testosterone secretion.

**Secondary female characteristics**

Female puberty is marked by:

*thelarche* (e.g., the onset of breast development),

*pubarche* (e.g., the appearance of pubic and axillary hair), and

*menarche* (i.e., the onset of menstruation).

Regular ovulation and fertility are attained about a year after menarche.

Estrogens maintain the development of related organs and female secondary sex characteristics.

Females have less body and facial hair and more fat beneath skin (i.e., a rounded appearance).

In females, the pelvic girdle enlarges, and the pelvic cavity is larger for wider hips.

## Female breast development

Estrogen and progesterone are required for breast development.

Female breast has a conical or *pendulous body* and a narrower *axillary tail* extending toward the armpit. *Breast* contains 15–24 lobules, each with a mammary duct.

*Nipple* is at the apex and is surrounded by a zone of darker skin as the *areola.*

Areola has small *areolar glands* that may appear as bumps around the nipple. They produce a secretion that prevents chafing and cracking of the skin in a nursing mother.

Mammary duct begins at the nipple and divides into ducts that end in alveoli (blind sacs).

Prolactin hormone is for *lactation* (milk production) to begin.

The feedback inhibition suppresses the prolactin production of estrogens and progesterone in the anterior pituitary during pregnancy.

Therefore, it takes a couple of days after delivering a baby for milk production to begin.

Before this, the breasts produce a watery, yellowish-white fluid (*colostrum*) similar to milk but containing more protein and less fat, rich in IgA antibodies that provide immunity to a newborn.

## Female hormone fluctuations

Breast cancer is a common cancer in females; women should have regular breast exams and mammograms as recommended.

At midlife, sexes go through a period of hormonal and physical changes of *climacteric.* This is marked by a decline in testosterone or estrogen secretion and a rise in FSH and LH secretion.

Female climacteric is accompanied by *menopause*, cessation of ovarian function and fertility.

Menopause is the age-related cessation of menstrual periods a decrease in the *number* of ovarian follicles and their *hypo-responsiveness to gonadotropins.*

Plasma estrogen levels decrease, resulting in high gonadotropin secretion.

A decrease in bone mass of osteoporosis occurs, hot flashes or the sudden dilation of arterioles, which increases body temperature and sweating.

## Male reproductive cycle

Hypothalamus controls the testes' sexual function through the secretion of GnRH.

GnRH stimulates the pituitary to produce the gonadotropic hormones FSH and LH in the anterior pituitary gland.

FSH promotes male spermatogenesis by stimulating primary spermatocytes to undergo meiosis I (forming secondary spermatocytes).

FSH enhances Sertoli cells (nurse cell that helps develop sperm) by causing them to bind to androgens effectively.

In males, LH is an *interstitial cell-stimulating hormone* (ICSH).

LH acts on Leydig's cells and stimulates testosterone production and secretion.

## Sperm production rate

*Sustentacular cells* of the seminiferous tubules release the hormone *inhibin*, which regulates the rate of sperm production and produces an androgen-binding protein, making the testes responsive to testosterone.

Hypothalamus-pituitary-testis system uses a *negative feedback* relationship that maintains a relatively *constant* production of sperm and testosterone.

Although hormone and gamete production is constant in males, this is not true for females.

## Neural control of sexual arousal

The penis contains vascular compartments and arteries.

Typically, these vessels are constricted with little blood, causing the penis to remain flaccid.

During sexual arousal, nervous reflexes cause an increase in the arterial blood flow to the penis.

Nerves of the penis converge on a pair of *dorsal nerves*, which lead via the *internal pudendal nerves* to the sacral plexus and then the spinal cord.

Penis receives *sympathetic, parasympathetic,* and *somatic motor nerve fibers.*

## Sexual excitation

Sexual excitation in higher brain centers, stimulation of mechanoreceptors in the penis, *inhibition of sympathetic fibers*, and release of *nitric oxide* contribute to arteriole dilation.

Dilation causes these compartments to engorge with blood at high pressure.

Increased blood flow fills and distends the erectile tissue, making the penis elongated and rigid, like an *erection.*

Erectile dysfunction is the inability to achieve an erection due to physiological or psychological causes.

Viagra and related products release *nitrous oxide* (NO) and block the breakdown of cGMP, a messenger involved in the relaxation of the arterial smooth muscle, promoting erection.

## Ejaculation

Stimulation of sympathetic nerves contracts the smooth muscles lining the ducts and discharges semen through the urethra.

Sphincter at base of the urinary bladder is closed so sperm cannot enter the bladder, and urine is not expelled.

*Ejaculation* is the expulsion of semen and is achieved at the peak of sexual arousal.

*Emission* is the *first phase of ejaculation.*

Nerve impulses from the spine trigger the epididymides and vasa deferentia to contract.

Subsequent motility causes the sperm to enter the ejaculatory duct.

Secretions are released from seminal vesicles, the prostate gland, and the bulbourethral glands.

A small amount of secretion from the bulbourethral glands may leak from the end of the penis to clean the urethra of acid, but it may contain sperm.

*Expulsion* is the *second phase of ejaculation.*

Rhythmical contractions at *penis base* and within the *urethral wall* expel the semen in spurts.

Rhythmical contractions are *myotonia* (muscle tenseness) release, a crucial sexual response.

Ejaculation lasts for a limited time, and the penis returns to a flaccid state following ejaculation.

*Refractory period* follows when stimulation does not result in an erection.

## Orgasm

*Orgasm* is a physiological and psychological sensation at the climax of sexual stimulation.

During orgasm, heart rate and blood pressure increase, and skeletal muscles contract.

*Clitoris* in females contains many sensory receptors as a sexually sensitive organ.

Female orgasm releases neuromuscular tension in the genital area, vagina, and uterus muscles.

*Notes for active learning*

## Hormonal Control of Ovarian Cycle

**Female ovarian cycle**

In longitudinal cross-section, an ovary has *cellular follicles*, each with an *oocyte* (egg).

A female is born with up to two million follicles.

The number is reduced to 300,000–400,000 by puberty, and a small number of follicles (about 400) fully mature.

As a follicle matures, it develops from a *primary* to a *secondary follicle* to a vesicular follicle (Graafian).

As oogenesis occurs, a secondary follicle contains a secondary oocyte pushed to one side of the fluid-filled cavity.

The vesicular follicle fills with fluid until the follicle wall balloons out on the surface and bursts, releasing a secondary oocyte surrounded by a *zona pellucida* and *follicular cells.*

During the *follicular phase*, FSH and LH stimulate primary follicles (containing primary oocytes) to grow and stimulate *theca cells*, expressing receptors for LH to produce androstenedione (androgen).

**Menstrual cycle**

*Endometrium* undergoes cyclic histological changes are the *menstrual cycle*, governed by the shifting hormonal secretions of the ovaries.

FSH, LH, estrogen, and progesterone are in a complex interaction to regulate menstruation.

An average 28-day uterine cycle is divided into four phases.

*Proliferative phase* is the mitotic rebuilding of tissue lost in the previous menstrual period and is primarily regulated by estrogens.

*Secretory phase* is regulated primarily by progesterone and consists of a thickening of the endometrium by secretions (not mitosis).

*Premenstrual phase* is ischemia and necrosis of the endometrium.

## Menstrual phase

*Menstrual phase* begins when endometrial tissue and blood are first discharged from the vagina and mark day 1 of a new cycle.

Decline in ovarian secretions of progesterone and estrogen triggers the menstrual phase.

Menstrual cycle begins on day 1, with *menstruation* (sloughing of endometrium with bleeding).

During days 1 to 5, low levels of estrogen and progesterone cause menstruation.

*Menstruation* is the periodic shedding of tissue and blood from the endometrium; this lining disintegrates, and the blood vessels rupture.

## Menses

*Menses* is the flow of blood and tissues discharged from the vagina.

FSH from the anterior pituitary stimulates the growth of a follicle in the ovary (the follicular phase), which secretes estrogen as it grows.

After day 5, the rising estrogen levels stimulate the uterus to grow a new inner lining.

Between days 6 and 13, increased production of estrogens by an ovarian follicle causes the endometrium to thicken and become vascular and glandular (proliferative phase).

By day 14, the endometrial lining is thick.

## LH surge

LH surge from anterior pituitary releases oocytes and some ovarian follicular cells (ovulation).

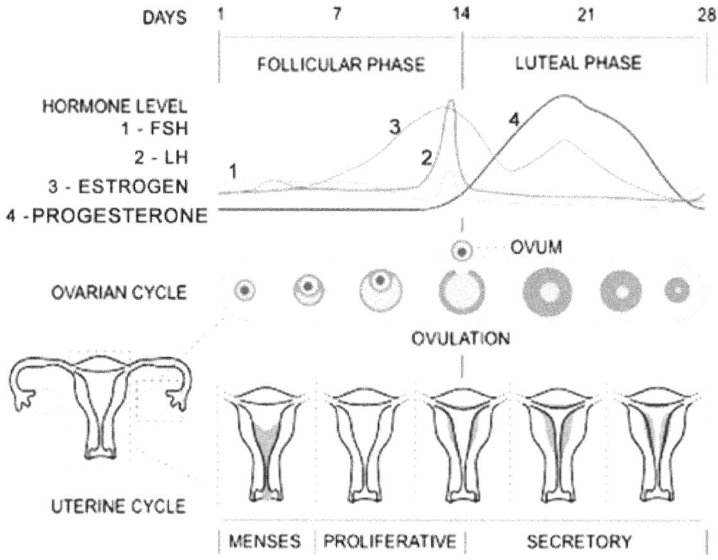


*Female ovarian and uterine cycles with associated hormone levels*

## Endometrium formation

As a response to LH, androgens are converted into estrogen by follicle-stimulating (FSH-induced) hormone by granulosa cells.

Estrogen leads to the thickening of the endometrium (uterine epithelium).

As estrogen levels rise, it exerts feedback control over the anterior pituitary secretion of FSH, causing the follicular phase to end.

As FSH decreases, the follicles cannot be maintained, and all but one follicle degenerates.

One dominant follicle (*Graafian follicle*) survives because it is:

> 1) hyperresponsive to FSH and can maintain itself under low FSH, and

> 2) sensitive to LH.

Estrogen levels rise, causing hypothalamus to secrete GnRH, causing a surge in LH secretion.

LH does not drop but shoots up (LH surge) because increased estrogen exerts *positive feedback* on the pituitary's LH-releasing mechanism.

LH spike triggers *ovulation,* while the remaining follicular cells become the *corpus luteum.*

*Notes for active learning*

## Hormonal Control of Female Reproductive Cycle

### Ovulation

*Ovulation* is the rupture of *vesicular follicle* discharging a 2° oocyte into the pelvic cavity.

Ovulation usually occurs on day 14 of the 28-day cycle.

*Secondary oocyte* completes a second meiotic cell division when fertilization occurs. Meanwhile, the follicle develops into the *corpus luteum* (promoted by LH), which secretes progesterone.

Days 15 through 28 have increased progesterone production by the corpus luteum, causing the endometrium to double in thickness.

Uterine glands mature, producing a thick mucoid secretion (secretory phase).

Endometrium is now prepared to receive an embryo.

Progesterone is the hormone responsible for maintaining the endometrium.

### Pregnancy maintains endometrium

Progesterone and estrogen levels decline *without pregnancy,* causing the lining to degrade and shed, initiating the next cycle.

With low levels of progesterone, the uterine lining begins to degenerate.

If pregnancy does not occur, the corpus luteum degenerates in about 14 days, and estrogen and progesterone levels recede.

Lack of estrogen and progesterone collapses vascular endometrium, leading to menstruation.

During menstruation, the anterior pituitary increases FSH production; a follicle begins maturation.

Ovarian cycle controls the *uterine cycle.*

### Hormonal response to fertilization

*Ovarian cycle* is under the control of gonadotropic hormones FSH and LH in females.

*Gonadotropic hormones* are not constant but are secreted at varying rates during the cycle.

*Luteal phase* in the ovary (corresponding to the uterus's secretory phase) is the second half of the ovarian cycle following ovulation.

Progesterone and estrogen from the corpus luteum inhibit the normal functioning of GnRH, which slows the production of FSH and LH.

Progesterone converts the endometrium into a secretory tissue full of glycogen and blood vessels, ready to receive a fertilized egg.

As progesterone levels in the blood rise, negative feedback decreases the anterior pituitary's secretion of LH, and the corpus luteum degenerates.

If fertilization does occur, the corpus luteum persists for about three months.

For example, postcoital contraceptives (e.g., Plan B) include progesterone antagonists that prevent progesterone from binding to receptors, leading to erosion of the endometrium.

When oral contraceptives (i.e., birth control pills) are used, synthetic progesterone and estrogen inhibit pituitary gonadotropin release, thereby preventing ovulation.

**Female and infertility**

*Infertility* is a condition where conception is difficult, but the individual *can conceive.*

*Sterility* means a person *cannot conceive*, regardless of medical or surgical interventions.

Major causes of female infertility:

      1) blocked oviducts

      2) failure to ovulate due to low body weight (<10-15%)

      3) *endometriosis*, the spread of uterine tissue beyond the uterus

      4) atypical hormone levels

Major causes of male sterility and infertility:

      1) low sperm count

      2) abnormal sperm from disease, radiation, chemical mutagens, or excessive heat near the testes

      3) atypical hormone levels

**Viviparity**

Mammals, including humans, are *viviparous*; the embryo develops in the female's body.

There are many forms of viviparity, but the most developed form is *placental viviparity*, where the mother constantly supplies the nutrients needed for development (e.g., placenta).

Human pregnancies are typically about *40 weeks* from the *first day* of the *last menstrual period*.

When the fetal brain matures, the hypothalamus causes the pituitary to stimulate the adrenal cortex to release androgens.

## Hormonal Control of Gestation and Birthing

### Pregnancy

Placental cells produce *human chorionic gonadotropin* (hCG), which maintains corpus luteum.

Human chorionic gonadotropin hormone is detected about 11 days after conception. Home pregnancy tests rely on the presence of HCG to confirm pregnancy.

In general, the level of HCG

> doubles every 48 hours during the first four weeks of pregnancy,

> doubles every 72 hours by 7-8 weeks, and

> peaks at 11-12 weeks of pregnancy.

### Corpus luteum

*Corpus luteum* produces *progesterone* and *estrogen*, maintaining uterus during first trimester.

HCG maintains the corpus luteum until the placenta produces its progesterone and estrogen, as the corpus luteum regresses.

*Progesterone* and *estrogen* have two effects at this stage:

> inhibit anterior pituitary, so no new follicles mature,

> maintain uterus lining, so the corpus luteum is not needed, eliminating menstrual cycles during pregnancy.

### Parturition

*Childbirth* (or *parturition*) is the completion of pregnancy and includes *labor* and *delivery*.

Placenta uses androgen precursors for estrogens stimulating prostaglandin and oxytocin.

Estrogen, prostaglandin, and oxytocin cause uterus to contract rhythmically and expel the fetus.

### Labor

*Labor* is a series of strong uterine contractions and has three stages:

> 1) cervix thins and dilates; amniotic sac ruptures and releases fluids;

> 2) rapid uterine contractions, followed by the birth of a newborn;

> 3) uterus contracts and expels the umbilical cord and placenta.

The cervix dilates, and the newborn moves through the vagina.

## Delivery

*Delivery* is the process of giving birth from the uterus or womb.

Following birth:

- oxygen is now supplied by breathing with functional lungs;

- switch from fetal circulation, which bypasses lungs and liver, to normal circulation (closing ducts and openings);

- nutrients now come from suckling rather than from the mother's blood.

## Breast development during pregnancy

Outside of pregnancy or lactation, the breast is primarily adipose and fibrous tissue with small traces of mammary glands.

Estrogen and progesterone secretion at the onset of puberty leads to breast enlargement due to the development of the duct system.

During pregnancy, the ducts grow, and branch and secretory *acini* (cell clusters) develop at the ends of the smallest branches.

During pregnancy, estrogen stimulates the secretion of prolactin and placental lactogen, stimulating the development of the glands (the alveoli) in breasts.

During pregnancy, milk secretion is *inhibited* by estrogen and progesterone.

The inhibitory effect stops after childbirth once the placenta is removed.

## Lactation

Lactation is the *secretion of milk* by the mammary glands.

Milk secretion is maintained by releasing prolactin from afferent input from nipple receptors to the hypothalamus during suckling.

*Dopamine* (inhibiting prolactin) secretion regulates *milk secretion*.

Milk ejection from the alveoli to ducts is stimulated by oxytocin, released by the suckling reflex.

Suckling inhibits hypothalamic-pituitary-ovarian hormones, which, in effect, blocks ovulation.

Breast is internally divided into lobes, each with a *lactiferous duct* conveying milk to the nipple.

Each duct expands into a *lactiferous sinus* just beneath the skin of the nipple.

*Acini* have contractile *myoepithelial cells* responding to oxytocin, causing milk flow in ducts.

*Notes for active learning*

*Notes for active learning*

# REVIEW

## Development

Page intentionally left blank

## Mechanisms of Reproduction

### Asexual reproduction

Reproduction may be *sexual* or *asexual*.

In sexual reproduction, *fertilization* initiates embryogenesis.

*Zygote* forms from the fusion of a sperm and egg.

*Asexual reproduction* does not involve fertilization, as only one parent is required.

Asexual reproduction is the primary form for single-celled organisms (e.g., bacteria, many protists, and fungi).

Rarely are animals exclusively asexual; animals that reproduce asexually combine sexual and asexual reproduction.

### Sexual reproduction

*Sexual reproduction* is when the *sperm* fertilizes the other parent's *ovum* (or *egg*).

Ovum and sperm are *gametes* produced by sexually mature organisms in *gametogenesis*.

Sexually reproducing animals use strategies to ensure their *gametes* (1N *germ line cells*) fuse.

*Germ line cells* are "*sex cells*" (i.e., *eggs and sperm*) that sexually reproducing organisms use to pass on their genomes to the next generation (i.e., parents to offspring).

Egg and sperm cells are *germ cells*, contrasting to "*body cells*" (*somatic cells*).

Ovum is available for fertilization by sperm *once per cycle during ovulation* in human females.

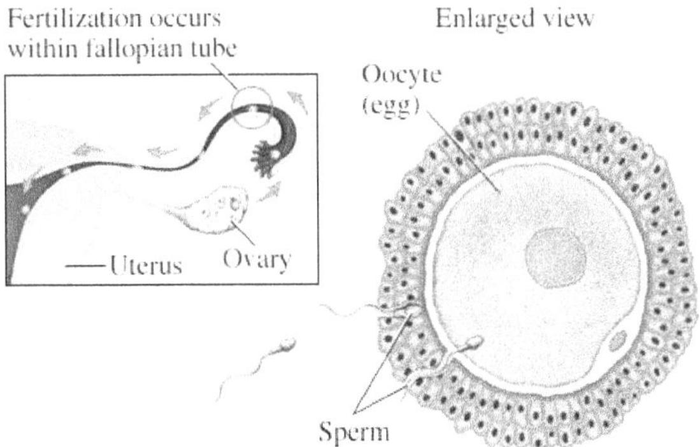


*Human reproductive structures and enlarged view of a fertilized ovum*

## External fertilization

Preserving the species requires passing genetic material; adults must produce fertile offspring.

Many animals are fertile during specific periods.

Sperm fertilizing an ovum may be *external* (e.g., in water) or *internal* (i.e., within an organism).

Animals using external fertilization (e.g., fish and amphibians) release thousands (or millions) of gametes (egg or sperm) since the chance of fertilization is low.

*Advantages of external fertilization* are greater genetic variation and low disease transmission.

Externally fertilizing animals may synchronize releasing gametes with environmental signals.

Tides, which fluctuate with the lunar cycle, are one such trigger.

## Internal fertilization

Most mammals (e.g., humans) use internal fertilization, and young develop within the mother's uterus until birth.

Animals using internal fertilization, including all terrestrial vertebrates, require one (or few) ova and are selective about reproductive partners.

*Disadvantage of internal fertilization* is that it reduces genetic variation and requires direct contact between males and females, promoting disease transmission.

Reptiles, birds, and monotreme mammals fertilize internally but subsequently lay eggs in which their young develop.

## Fertilization mechanisms

*Ovary* releases an *ovum*, swept by ciliary cells into the uterine tube (*oviduct* or *Fallopian tube*) during *ovulation*.

During intercourse, millions of sperm are ejaculated and move through the vagina, uterus, and uterine tube towards the ovum using their *flagella*, whip-like tails.

*Temperature* and *chemical signals* direct sperm toward the ovum.

*Rhythmic propulsions* of the vagina and uterus aid sperm in their journey.

Most sperms die before reaching the ovum from depleted energy (ATP) and an acidic vagina.

Sperm secretes proteins that bind to receptors on the *glycoprotein layer* surrounding the plasma membrane of a same-species ovum, preventing cross-species fertilization.

Glycoprotein layer is the *vitelline layer* or *zona pellucida* in mammals.

*Acrosome* in sperm head digest *zona pellucida*, releasing *hydrolytic enzymes* upon binding.

## Gametes fuse

Sperm cells reaching the ovum are viable and have the potential to fertilize it.

Sperm cells attach to the ovum, but only one should penetrate the ovum.

Fertilization requires gametes to fuse haploid (1N) nuclei as pronuclei.

After fusion, the sperm body follows the haploid (1N) pronuclei into the ovum.

Two pronuclei form a diploid nucleus containing the zygote's unique genome.

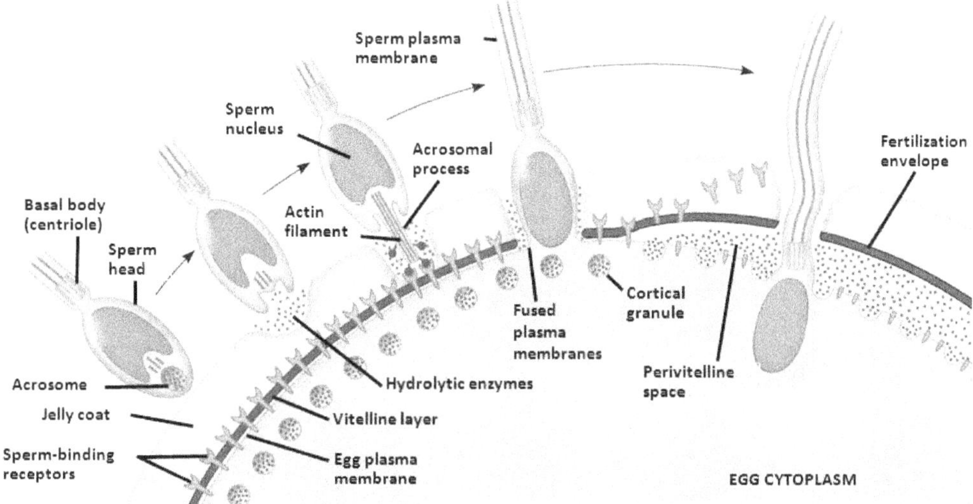


*Fertilization of an egg with time series (clockwise) initiated by sperm binding*

*Notes for active learning*

## Mechanism to Avoid Polyspermy

### Polyspermy

*Polyspermy* is when more than one sperm penetrates the ovum.

Usually, a fertilized ovum has 2 copies (2N or *diploid*) of each chromosome.

An ovum affected with polyspermy has 3 or more copies, forming a non-typical zygote.

For example, Down's Syndrome is trisomy 21.

Preventing polyspermy depends on specific changes when sperm binds and fuses to an ovum.

The ovum is vulnerable to sperm attempting to fertilize it, so immediate action is taken to avoid polyspermy by two mechanisms:

1) *fast block* and

2) *slow block*

### Fast block to polysperm y

In marine invertebrates (e.g., sea urchins), fertilization triggers an influx of sodium ions into the ovum, which causes its membrane potential to depolarize rapidly.

Sperm cells cannot bind to the positively charged ovum, so *fast block* prevents polyspermy.

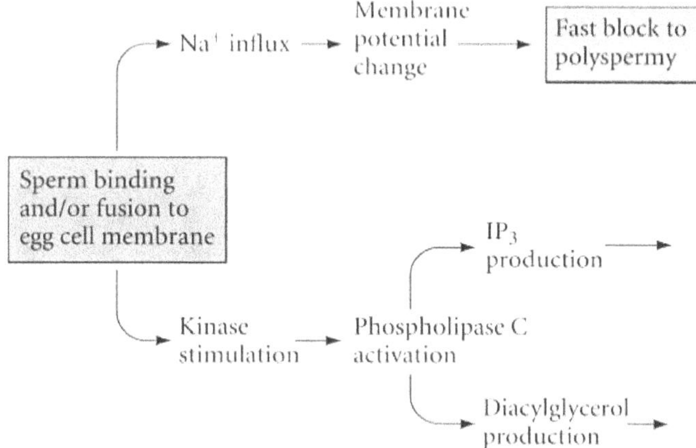


*Fast block to polyspermy involves a change to the membrane potential*
*that inhibits additional sperm from penetrating the fertilized egg*

**Slow block to polyspermy**

*Slow block* is a gradual but longer-term process.

Mammals use slow blocks to polyspermy and do *not combine* fast and slow blocks as some species (e.g., sea urchins).

Many species other than mammals utilize *both* fast and slow blocks to polyspermy.

*Fast block* is effective and transient; another countermeasure to polyspermy is often needed.

*Slow block* is initiated upon fertilization, as the ovum secretes various hormones to prevent it from being overwhelmed by the hundreds of sperm attempting fertilization.

In most species using this technique, the slow block is the *cortical reaction.*

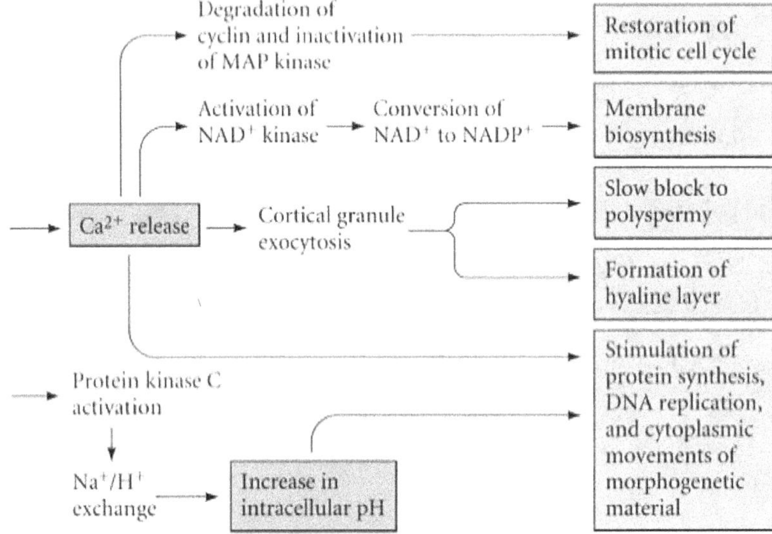


*Slow block to polyspermy uses cortical granules to create a physical barrier around the zygote*

**Cortical reaction**

*Cortical reaction* is calcium-dependent exocytosis of secretory granules released into the *perivitelline space* at fertilization, modifying the *zone pellucida* to prevent polyspermy.

Proteases (*proteins digesting proteins*) lift the *vitelline layer* as *hyaluronic acid* pushes the vitelline layer away from the ovum's surface.

This separation forms a barrier and prevents entrance by other sperm.

*Hyalin* and *peroxidases harden* the *fertilization envelope* and *inactivate sperm-binding sites*, preventing fusion by other sperm.

## Zona reaction

*Vitelline layer* is the *fertilization envelope.*

Vitelline layer is the *zona pellucida,* so the *cortical reaction* is the *zona reaction* in mammals.

*Zona pellucida hardens,* and its binding sites become inactivated.

*Zona reaction* of mammals is less effective than the cortical reaction in other species.

Zona reaction and cortical granules inactivate ZP2 and ZP3 sperm receptors on *zona pellucida.*

## Sperm pronucleus and centrioles

Evidence suggests that sperm mitochondria enter but are destroyed by the ovum.

Sperm centrioles and pronucleus (i.e., gametic nuclei) enter the ovum.

Sperm contributes more than its *pronucleus* to the zygote.

Centrioles replicate and assemble into centrosomes, replicating the first mitotic spindle assembly in the zygote.

Extra mitotic spindles have been observed in polyspermy when several sperm cells contribute centrioles to the ovum.

*Notes for active learning*

## Fertilization Cascade

### Zygote as a fertilized egg

Zygote is propelled by *ciliary movement* through the uterine tube (*oviduct* or *Fallopian tube*) into the uterus; it begins cell divisions as *cleavage*.

*Development* describes the changes in the life cycle of an organism.

*Embryogenesis* begins the process of growth and development of an embryo.

*Diploid* (2N) fertilized zygote begins mitotic division as an *embryo*.

### Indeterminate *vs.* determinate cleavage

*Cleavage* is cell division *without growth*; *blastomeres* (cells) are smaller with each division.

Cleavage is *indeterminate* or *determinate*.

> *Indeterminate cleavage*: blastomeres individually complete development if separated.

> *Determinate cleavage*: blastomeres do not develop if separated; each is necessary.

*Determinate cleavage* is typical for *protostomes*, a superphylum of animals (e.g., annelids, arthropods, nematodes, platyhelminths, rotifers, and mollusks).

### Mosaic *vs.* regulative cleavage

*Protostomes* are the counterpart to *deuterostomes*, well-known as *chordates* (i.e., *vertebrates*) and *echinoderms* (e.g., *sea urchins*).

*Deuterostomes* typically display *indeterminate cleavage*, which gives rise to identical twins.

*Identical twins* are *monozygotic* twins from a single zygote division into two separate embryos.

*Fraternal twins* are two separately fertilized ova that independently develop in the uterus.

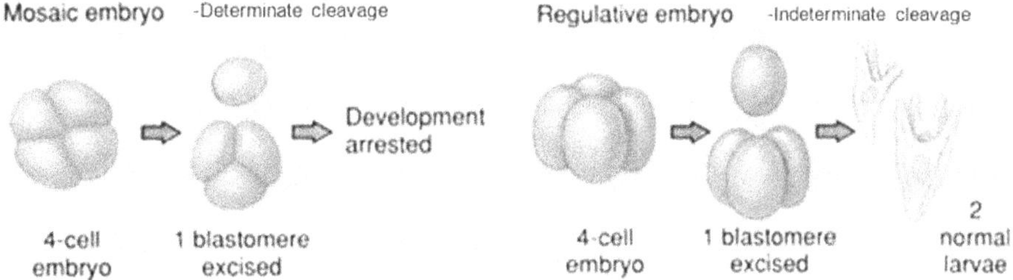


*Mosaic embryo with determinate cleavage vs. regulative embryo with indeterminate cleavage*

## Egg polarity

*Embryo polarity* is when an ovum has an *animal pole* (upper) and *vegetal pole* (lower hemisphere) in many species.

*Vegetal pole* contains larger cells filled with *yolk.*

*Yolk* is a nutritious substance that feeds the embryo.

Yolk is denser than the cytoplasm, which causes it to settle at the bottom.

Yolk differentiates into extraembryonic membranes, protecting and nourishing mammalian embryos.

*Animal pole* is smaller, more rapidly dividing cells and forms the *embryo proper.*

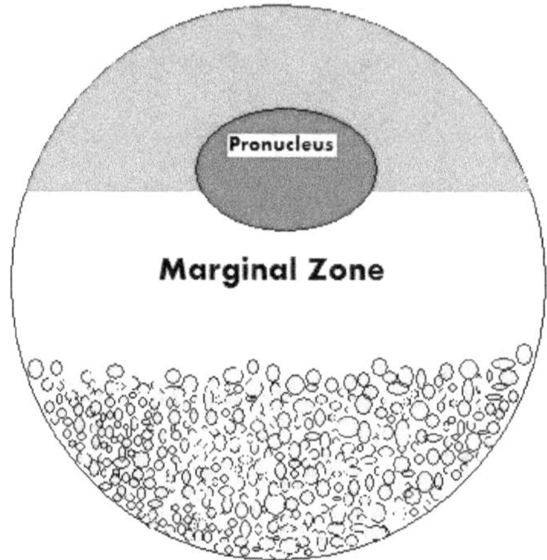


*Animal pole is on the upper portion of the egg while vegetal pole is the lower portion of the egg; pronucleus is at the boundary above marginal zone*

*Archenteron* is the center cavity formed by gastrulation, and its opening is the *blastopore.*

## Blastopore fate defines protostomes and deuterostomes

*Fate of the blastopore* defines deuterostomes and protostomes.

*Protostome* (*first mouth*): blastopores first form the *mouth*, and *anus* develops second.

*Deuterostome* (*second mouth*): blastopores first form *anus*, and *mouth* develops second.

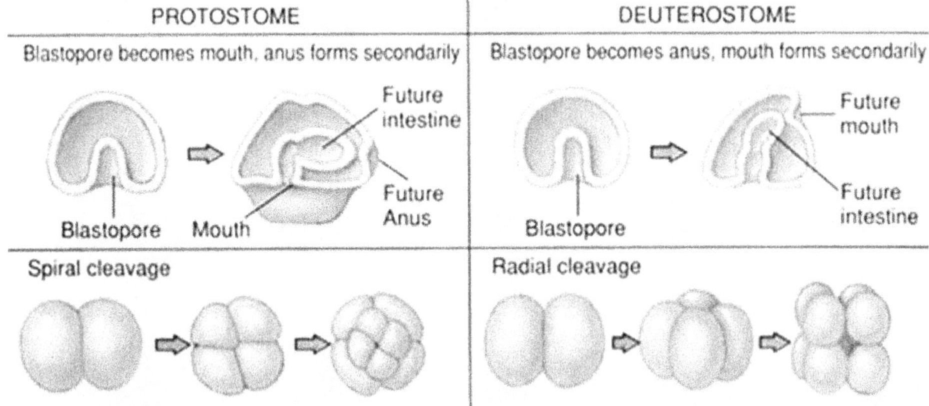


*Protostome vs. deuterostome and spiral vs. radial cleavage comparing blastopore fate*

## Radial *vs.* spiral cleavage

First zygote cleavage is *polar*, dividing the ovum into two along the *vegetal-animal axis*.

Subsequent cleavages are *perpendicular* to the vegetal-animal axis, which is *equatorial*.

*Radial cleavage* alternates between polar and equatorial cleavage.

Radial cleavage is typical for *deuterostomes*.

*Spiral cleavage* is typical for *protostomes*.

Yolk volume in the zygote influences the manner of cleavage.

For example, species laying external eggs (e.g., birds, some fish, reptiles, and most insects) have much yolk, leading to *meroblastic* (i.e., *incomplete*) cleavage.

Mammals, worms, insects, and fish have less yolk.

Less yolk is typical for *holoblastic cleavage* (i.e., *complete cleavage*).

*Spiral and determinate vs. radial intermediate cleavage at the eight-cell stage*

*Notes for active learning*

## Embryogenesis

### Morula and blastula formation

A few days after fertilization, cleavage creates a *morula*, a solid ball of cells.

By the fifth day, the morula is transformed into a hollow *blastula* (*blastocyst* in humans).

*Blastocyst* has an outer ring of cells as a *trophoblast* and an inner cell mass, the *embryoblast*.

Blastocyst is formed as *blastomeres migrate to the outside* of the morula, leaving behind the *blastocoel*, a *fluid-filled cavity.*

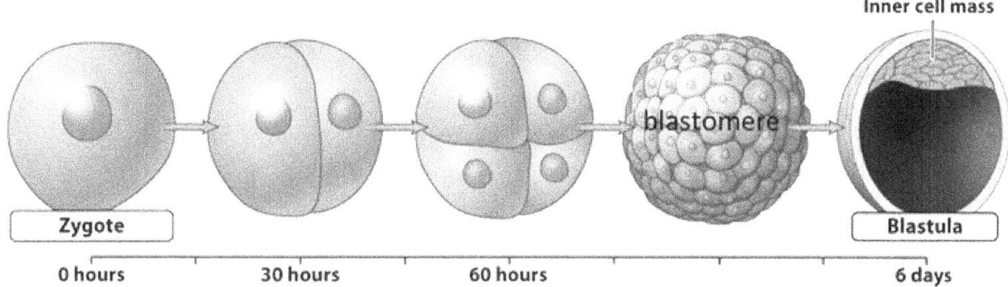


*Fertilized egg is a zygote that proceeds by cytoplasmic cleavage to form a morula, becoming a blastula*

### Blastocyst implantation

*Trophoblast* accomplishes blastocyst implantation by embedding into the *endometrium.*

*Endometrium* is the nourishing epithelium of the uterus.

Trophoblast releases *human chorionic gonadotropin* (hCG) upon implantation.

*Human chorionic gonadotropin* maintains estrogen and progesterone production from the *ovary's corpus luteum*, maintaining the endometrium.

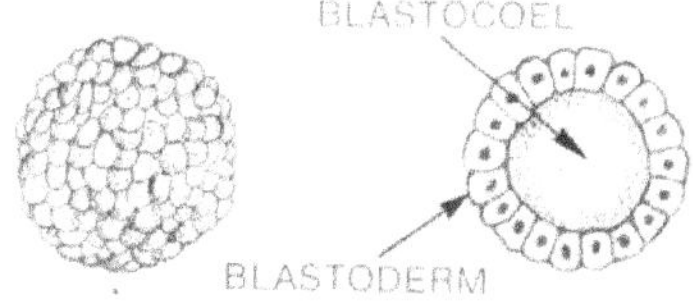


*Blastula with a surface (left) and cross-sectional view of outer blastoderms and inner blastocoel*

## Yolk

*Yolk amount* affects the first three stages of development (cleavage, blastulation, gastrulation).

For example, the lancelet and frog develop quickly in water and travel into the larvae to feed.

In contrast, the chick provides a significant amount of yolk inside a hard shell, and its development continues until it can exist independently.

*Human initial stages* resemble the chick due to our shared evolutionary history.

Frog embryo cells at the animal pole have little yolk; cells at the vegetal pole have more.

Frog blastocoel forms at the *animal pole.*

Yolk causes cells to cleave more slowly, so cells at the animal pole are smaller.

For example, chick cell cleavage is *incomplete*; only those cells lying on the yolk cleave and spread over the yolk surface.

Chick cleavage is contrasted with the *ball-like morula* (*solid ball of cells*) of lancelets.

## Archenteron and blastopore

Yolk cells do *not* participate in gastrulation and do *not* invaginate; instead, when animal pole cells evaginate from above, slit-like *blastopore* forms.

Other pole cells move down over the yolk; the blastopore becomes rounded.

Yolk cells temporarily remaining in the region form a *yolk plug.*

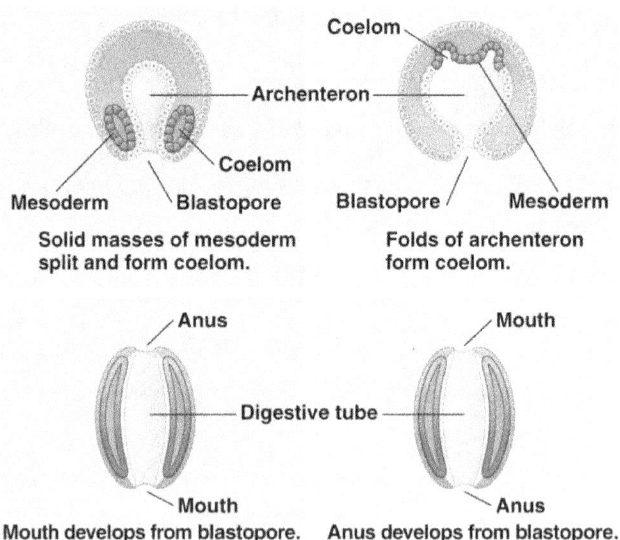


*Protostome vs. deuterostome comparing blastopore formation and mouth vs. anus sequence*

   

## Cell Migration

### Dorsal lip

*Dorsal lip* cells of blastopore migrate between the ectoderm and endoderm, forming *mesoderm*.

Later, the *coelom* is created by splitting off from the mesoderm.

For example, a chick's blastocoel is created when cells lift from the yolk and create space between cells and yolk; so much yolk that endoderm formation does not occur by invagination.

Upper layer of cells differentiates into *ectoderm*, and lower layer differentiates into *endoderm*.

*Mesoderm* arises by an invagination of cells along the edges of the longitudinal furrow in the embryo midline, named the *primitive streak*.

Later, the newly formed mesoderm splits to form the *coelomic cavity*.

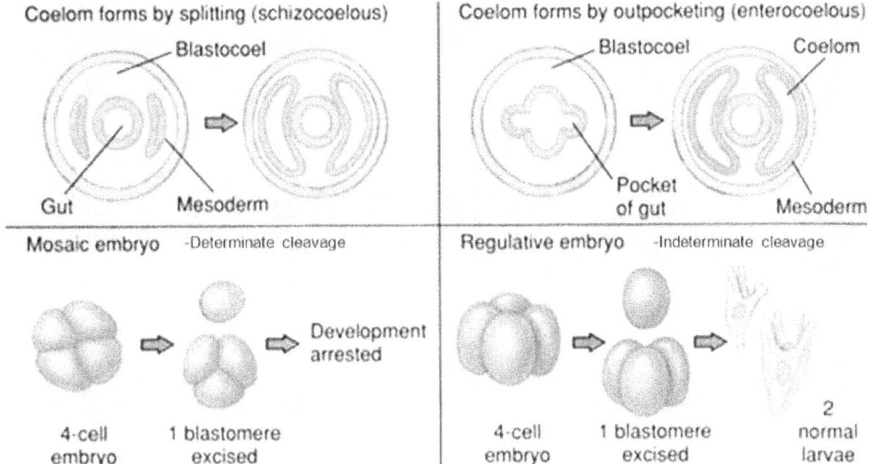


*Protostomes (left) form coelom by splitting compared to deuterostomes that use out pocketing*

## Gastrulation

*Gastrulation* is a series of orchestrated cell movements modulated by complex biochemical cell-signaling pathways; gastrulation varies by species.

After fertilization (3 weeks in humans), *gastrulation* is when an invagination of cells into the blastocoel forms the *primary germ layers*.

*Primary germ layers* are the initial cell layers from which *all* body tissues develop.

Vertebrates and higher animals are *triploblastic* with three primary germ layers.

Primitive animals (cnidarians and sponges) are *diploblastic*, with two germ layers.

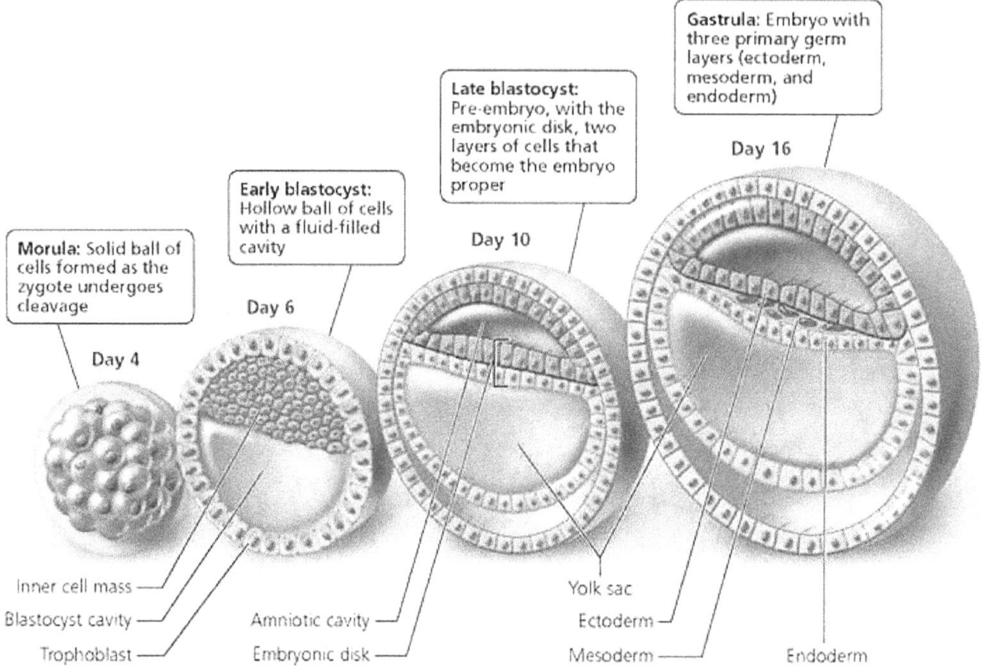


*Morula, blastula, and gastrula with the formation of three primary germ layers*

## First cell movements

In humans (and other mammals), the inner cell mass flattens into the embryonic disc before gastrulation, divided into two layers: the *epiblast* and the *hypoblast*.

Gastrulation is marked by a *primitive streak,* a line of cells along the *embryo's midline.*

*Epiblast ingresses* along with the primitive streak, pushing the *hypoblast.*

*Blastopore* is this first ingression.

As invagination continues, the blastopore deepens to form the *archenteron*, the *primitive gut.*

*Hypoblast* becomes the amnion, one of the extraembryonic membranes.

## Primary Germ Layers

### Gastrulation forms primary germ layers

Germ layers form during gastrulation when a blastula (hollow ball of cells) begins to differentiate into more-specialized cells that become layered across the developing embryo.

Each germ layer gives rise to specific tissue types.

*Epiblast cells* that migrate inwards differentiate into the *mesoderm* (middle germ layer).

*Diploblasts* (sponges, cnidarian) develop *mesoglea*, a non-cellular layer, instead of a mesoderm.

*Epiblastic cells* that continue to invaginate the blastocyst become the *endoderm* (*inner* germ layer), while those that remain outside are the *ectoderm* (*outer* germ layer).

### Germ Layers

After gastrulation, the *germ layers* develop into various structures and organs.

Germ layers represent *early lineage-specific* (i.e., multipotent) *stem cells* (i.e., cells destined as specific tissue, such as muscle or blood) in embryonic development.

Germ layers differentiate into various organs and tissues; they are valuable in studying development and stem cell research.

Scientists reliably produce specific cell types from stem cells and induce pluripotent stem cells (genetically reprogrammed adult cells), which yield novel cell-based therapies.

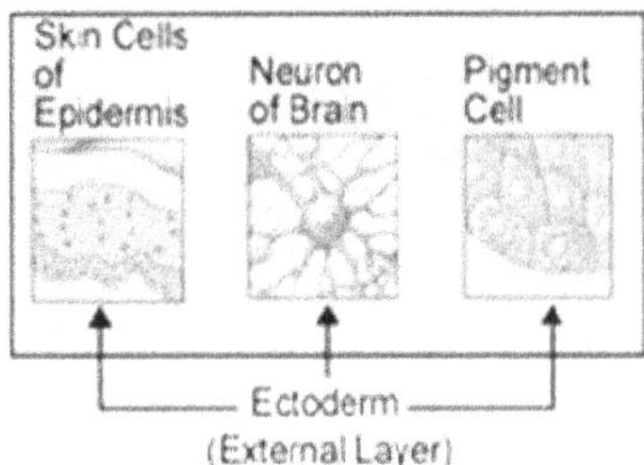


*Ectoderm is the outermost layer and gives rise to skin, nerve tissue, and epidermal skin cells*

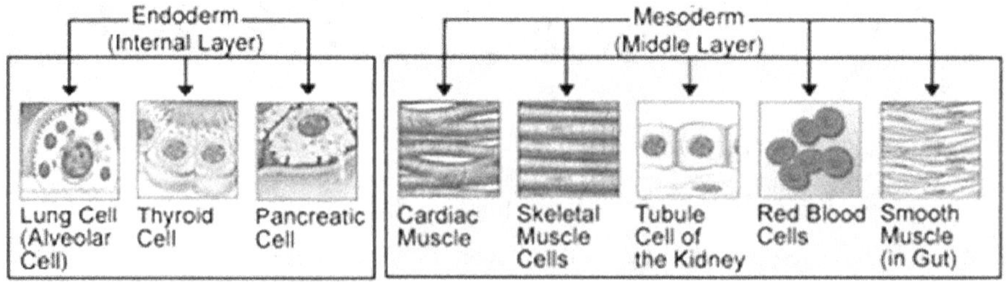



*Primary germ layers of endoderm and mesoderm and representative tissue types formed from each germ cell layer*

## Ectoderm, endoderm and mesoderm

*Ectoderm* forms specific outer linings; the epidermis (i.e., outermost skin) and hair.

*Mammary glands*, *central* (CNS), and *peripheral nervous systems* (PNS) are ectodermal.

*Endoderm* is the innermost of the three germ layers and gives rise to specific organs: the colon, stomach, intestines, lungs, liver, and pancreas.

Endodermal cells form many internal linings, including the lining of most of the gastrointestinal tract, the lungs, the liver, the pancreas, and glands that open into the gastrointestinal tract and other organs (e.g., the *upper urogenital tract* and *vagina*).

*Mesoderm* lies between endoderm and ectoderm and gives rise to all other tissues, including skin dermis, heart, muscle, urogenital systems, bones, and bone marrow (e.g., blood).

## Ectoderm

*Ectoderm* gives rise to:

> nervous system,
>
> skin epidermis,
>
> epithelial lining of the mouth and rectum,
>
> glands, nails, and
>
> hair, skin glands, and epidermal derivatives.

**Mesoderm**

*Mesoderm* forms the excretory and reproductive system, the heart, blood vessels and blood cells, and linings of all body cavities (coeloms).

Portions of mesoderm develop into connective, muscular, and skeletal body portions.

Mesoderm is the germ layer that *distinguishes evolutionarily higher life forms* (i.e., bilateral symmetry) from lower life forms (i.e., radial body symmetry).

Mesoderm allows more highly evolved organisms to have an internal body cavity to protect organs, bathing them in fluids and supporting them with connective tissue.

**Somites**

During neurulation, *midline mesoderm cells* not contributing to the notochord formation become two longitudinal masses of tissue as *somites*, a process of *somitogenesis.*

*Vertebrate somites* form dermis, skeletal muscle, tendons, cartilage, and axial skeletal bones.

**Endoderm**

*Endoderm* gives rise to:

digestive tract,

respiratory tract, and excretory tract epithelial linings,

organs and glands associated with the digestive and respiratory systems.

**Structures from primary germ layers**

| Ectoderm | Mesoderm | Endoderm |
|---|---|---|
| Epidermis<br><br>Nervous system<br><br>Epithelial linings: mouth and rectum<br><br>Glands: adrenal medulla, pineal, and pituitary | Dermis<br><br>Skeletal and muscular, excretory, reproductive, circulatory, and lymphatic systems<br><br>Epithelial linings: coelom<br><br>Glands: adrenal cortex | Epithelial linings: respiratory, digestive, and excretory systems<br><br>Glands: pancreas, thymus, thyroid, and parathyroid<br><br>Liver |

*Notes for active learning*

## Neurulation

### Notochord

*Mesoderm cells* of gastrulation coalesce along the central axis, forming a dorsal *notochord.*

*Notochord* is a stiff rod supporting lower chordates.

*Vertebral column* replaces the notochord in amphibians, birds, and mammals.

Ectoderm just above the notochord develops into the nervous system in *neurulation.*

Ectoderm cells on the dorsal surface of the embryo thicken, forming the *neural plate.*

### Neural grove and neural folds

*Neural groove* develops down the midline of the neural plate.

On the sides of the neural groove, *neural folds* move upward and fuse, forming the *neural tube,* which becomes the central nervous system.

At this stage, the embryo is a *neurula.*

### Neural tube

*Neural crest* is the ectoderm where the two *neural folds* close to form the *neural tube.*

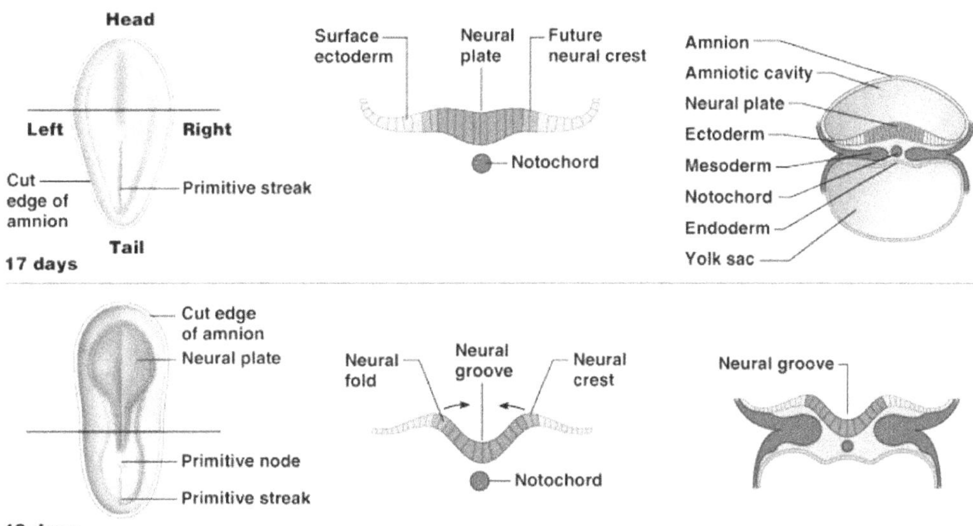


*Neural crest is ectoderm that forms during neurulation*

## Neural crest

Neural crest cells *migrate and differentiate* into:

> neurons,
>
> glial cells of the peripheral nervous system,
>
> melanocytes of the epidermis,
>
> epinephrine-producing cells of the adrenal gland,
>
> portions of the heart, and
>
> connective tissue of the head.

*Neural crest* (derived from ectoderm) is the *fourth germ layer* because of its importance.

## Neural tube formation

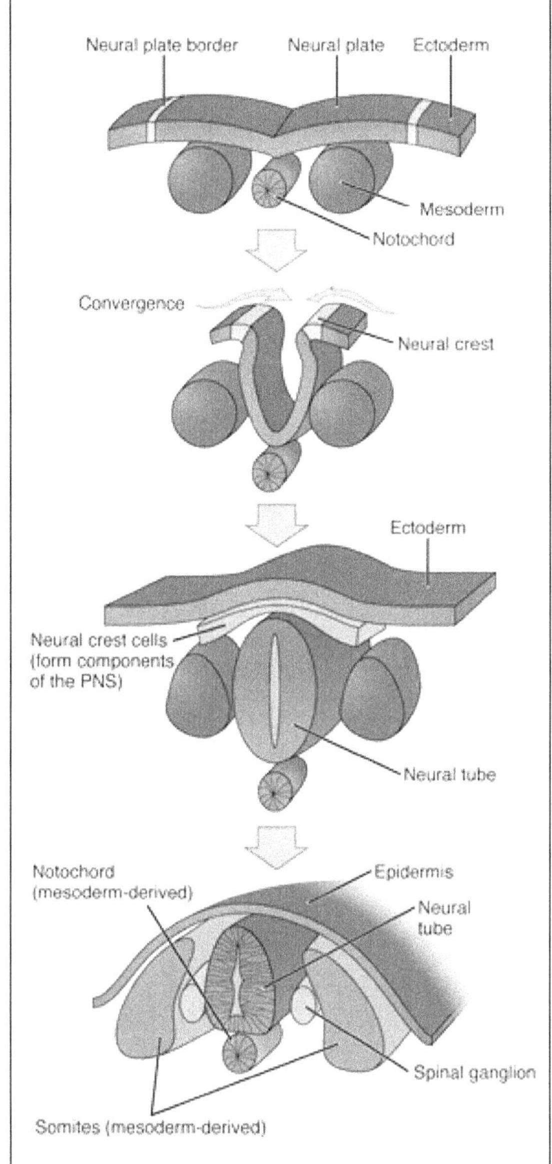


1 – *Neuroectodermal tissues* differentiate from ectoderm and thicken into neural plates. *Neural plate borders* separate the ectoderm from the *neural plate*.

2 – *Neural plate* bends dorsally, with two ends joining at the neural plate borders, now the *neural crest*.

3 – *Neural tube closure* disconnects the neural crest from the epidermis. *Neural crest* cells differentiate to form most of the *peripheral nervous system* (PNS).

4 – *Notochord degenerates* and persists as nucleus pulposus of *intervertebral discs*. Mesoderm cells differentiate into *somites, axial skeleton*, and *skeletal muscle precursors*.

*Notes for active learning*

## Mechanisms of Development

### Organogenesis

*Development* is a changing life cycle with *growth, differentiation,* and *morphogenesis.*

Most animals undergo the same embryonic stages:

zygote, morula, blastula, and gastrula

*Gastrulation* transitions to *neurulation,* the first organogenesis event with organ development.

*Neurulation* is when the embryo transitions from the embryonic stage to a *fetus.*

*Organogenesis* relates to *morphogenesis,* developing an organism's body plan and shape.

### Cell specialization

*Specialization* refers to the cell type when they have a specific function.

For example, epidermal cells produce *keratin* to protect the skin against abrasion, myocytes produce *actin and myosin* to make muscles contract, and neurons produce *neurotransmitters* that transmit electrochemical impulses.

*Stem cells* are undifferentiated cells that have yet to specialize.

As an embryo develops, its stem cells differentiate to give rise to specific tissues.

*Cell specialization* is achieved in two stages:

*determination* and

*differentiation*

### Determination

*Cytoplasmic influences* cannot change the final form of a determined cell.

Cytoplasmic influences narrow with successive cell divisions; complete determination usually occurs late in cell specialization.

Determination begins with specification (reversible) and ends with an irreversible commitment (differentiated) to a cell type.

*Determination* is when cells commit to a cell type as *specified* and *determined.*

## Specified cells

*Specified cells* transplanted into different body tissue alter their differentiation pathway to match the recipient tissue.

In this regard, specification is *reversible.*

*Determined* is when cells have restrictive potential and commit to specific cell types.

Determined cells are *irreversibly committed* to their differentiation pathway.

*Determined cells* do *not* adapt to a different environment by altering their differentiation.

## Differentiation

*Cellular differentiation* is when cells become specialized in their structure and function.

*Determined* and *differentiated* are sometimes used interchangeably.

*Differentiation* is a *morphological* and *biochemical* process of a cell's final form and function.

*Differentiation* is an action, while *specialization* is a state of being.

## Morphogenesis

Differentiation is an essential component of *morphogenesis.*

*Morphogenesis* is when the overall form of the organism and its body parts are shaped.

Morphogenesis includes early cell movements and later *pattern formation,* the process by which cells assume distinct functions in specific body parts.

As cells differentiate, functions change and interact, eventually organizing into body tissues.

Cells can be traced during differentiation in a developmental lineage map by *fate mapping.*

## Differential gene expression

Each body cell contains a complete set of chromosomes and, therefore, all genetic information required to perform the functions of any cell.

Cells have the same genome but differ in gene expression (i.e., transcription) and degree.

Cells do not differentiate because they receive different genes, or genes are lost as cells divide.

Instead, cell specialization is accomplished via *differential gene expression.*

## Cytoplasmic segregation

*Cytoplasmic segregation* and *induction* are two crucial developmental mechanisms of differential gene expression.

The foundation of cell specialization in an embryo is laid before fertilization.

Ova contains mRNA and proteins as *maternal determinants* that influence development.

Blastomeres receive different concentrations of maternal determinants, leading to *differential gene expression.*

*Cytoplasmic segregation* concentrates maternal determinants within ova as cleavage occurs for differential gene expression.

For example, the cytoplasm of a frog zygote is not uniform.

After the first division of the frog zygote, only daughter cells receiving portions of the *gray crescent* develop into a complete embryo.

## Induction

*Induction* is a more common mechanism of differential gene expression, in which one group of cells influences development by changing the behavior of an adjacent group of cells.

Hans Spemann (1869-1941) received the Nobel Prize in 1935 for *embryonic induction* and found that chemical signals within the *gray crescent* activate *genes regulating development.*

*Notes for active learning*

## Tissue Formation

### Four tissue types

Four types of animal tissue are epithelial, connective, nervous, and muscle.

Each tissue type contains specialized cells and an extracellular matrix secreted by cells.

### Epithelial tissue

*Epithelial tissue*s may be multilayered or one cell thick and usually function in secretion, absorption, or protection.

Epithelial tissues are in the lining of glands, organs, body cavities, blood vessels, skin, and mucous membranes.

Connective tissue

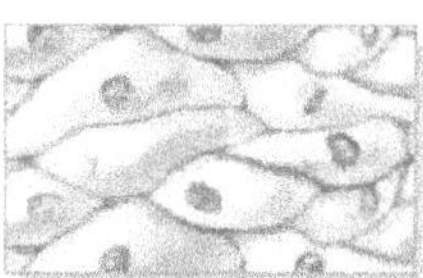

Epithelial tissue

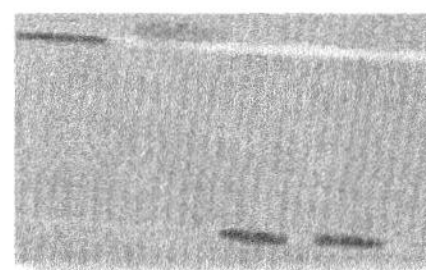

Muscle tissue

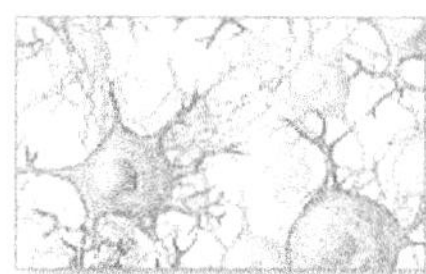

Nervous tissue

Various epithelia are derived initially from all three germ layers.

### Connective tissue

*Connective tissue*s tether tissues and organs but can have structural and support functions.

Cartilage, bone, blood, and adipose tissue (fat) are connective tissues, the most versatile and widespread of body tissues.

Proper connective tissues divide into *loose* and *dense connective tissues* divided into subtypes.

These tissue types are derived initially from *mesenchyme,* a portion of the mesoderm.

*Mesenchyme* is a loosely organized, mainly mesodermal embryonic tissue which develops into *connective and skeletal tissues*, including *blood and lymph*.

### Nervous tissue

*Nervous tissue* includes the *neurons* (nerve cells) and *glial cells* of the nervous system.

*Neurons* transmit electrical signals, while glial cells protect and support the neurons.

*Glial cells* provide physical and chemical support to neurons and maintain their environment.

Glial cells are in the central nervous system (CNS) and peripheral nervous system (PNS).

Central and peripheral nervous tissue arises from the *ectoderm during neurulation*.

**Muscle tissue**

*Muscles are* contractile tissue divided into *skeletal, smooth,* and *cardiac* muscles.

*Skeletal muscle* tissue is under voluntary control and attached to the skeletal system by *tendons*.

*Smooth muscle* is involuntary in walls of hollow organs (e.g., uterus, stomach, blood vessels).

*Cardiac muscle* is under involuntary control but only in the heart, which creates powerful contractions that move blood throughout the body.

All three muscle types are derived initially from *mesoderm*.

## Pregnancy Cascade

### Pregnancy stages

*Human gestation* is nine months; calculated by adding 280 days to start of last menstrual cycle.

5% of infants arrive on their forecasted birth date due to several complicated variables.

*Embryonic development* is months 1 and 2 with zygote *cleavage* and initial organ development (*organogenesis*).

*Fetal development* is months 3 through 9 when organ systems grow and increase in size and weight by a factor of 600.

### Implantation

After fertilization, *blastocyte implantation* in the endometrium occurs at about week one.

Upon implantation, the trophoblast secretes *human chorionic gonadotropin* (hCG) to stimulate the *corpus luteum* to maintain the endometrial lining.

As weeks progress, the inner cell mass detaches from the trophoblast to become the *embryonic disc*, while the yolk sac forms below.

### Yolk sac

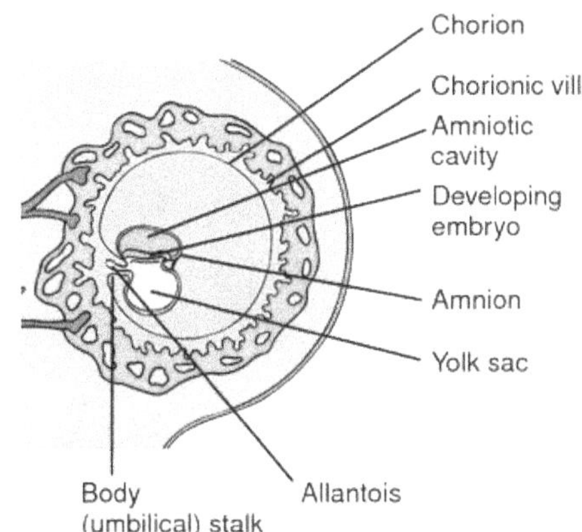


*Yolk sac* is a membranous sac attached to embryo.

*Extraembryonic membranes* lie outside the embryo, protecting and nourishing it and, later, the fetus.

Embryos developing in water receive oxygen from water, and wastes float away.

Surrounding water prevents desiccation and is a protective cushion.

Evolution of extraembryonic membranes in reptiles made development on land possible.

*Notes for active learning*

## Extraembryonic Membranes

### Membrane modifications

*Extraembryonic membranes* perform these functions for an embryo on land.

Membrane modifications depend if the organism undergoes internal or external development.

*Yolk sac* nourishes the developing embryo in birds and reptiles.

*Umbilical cord* and *placenta* nourish the embryo in placental mammals.

*Yolk sac* is empty of yolk and instead functions as an early blood supply for the embryo before becoming part of the *primitive gut* later in gestation.

Therefore, it is an *umbilical vesicle* to distinguish it from the yolk sacs of egg-laying animals.

### Amnion extraembryonic membrane

Yolk sac forms and a second extraembryonic membrane of the *amnion* encloses the embryo.

*Amnion* surrounds the *amniotic cavity*.

*Amniotic cavity* contains protective amniotic fluid bathing the developing embryo.

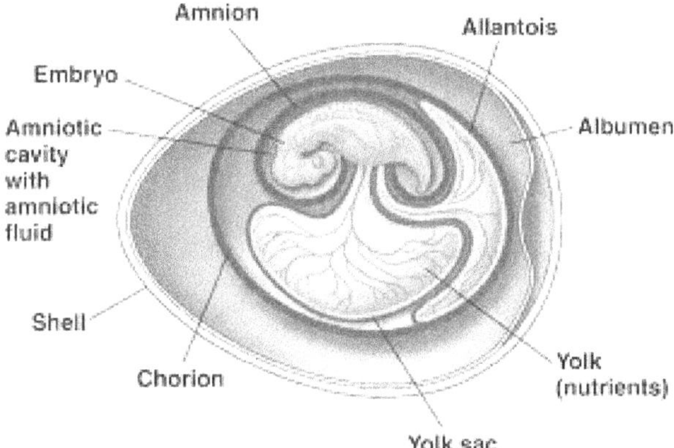


*Egg-laying animals have large yolk sacs with nutrients needed during embryogenesis*

Chordates, whether externally or internally, develop in water or *amniotic fluid*.

*Amniotic fluid* is a *buffer* for mechanical and chemical disturbances and *thermal regulation*.

*Amniocentesis* is a prenatal test when the amniotic fluid is sampled for genetic analysis.

## Chorion extraembryonic membrane

*Chorion* is an extraembryonic membrane that, in birds and reptiles, lies next to the shell and performs gas exchange.

In placental mammals, the chorion implants into the endometrium and projects treelike extensions called *chorionic villi* into the maternal tissues.

Maternal blood circulates these villi, so molecules in fetal and maternal blood are exchanged.

$CO_2$ and waste move from the fetus, and $O_2$ and nutrients flow from the maternal side.

## Placenta

As the embryo develops, the interface of *chorionic villi* and *uterine tissue* becomes the *placenta.*

*Placental formation* begins about week 1 after the embryo is implanted in the endometrium.

*Placenta* is the life support structure of the developing embryo.

Placenta provides *oxygen* and *nutrients* while *removing wastes* (e.g., $CO_2$).

## Umbilical cord

In humans, maternal blood contacts the chorion; gas, nutrients, and wastes cross the chorion without mixing maternal and fetal blood.

Placenta facilitates gas and nutrient exchange but not fluid exchange.

*Umbilical cord* stretches between the placenta and fetus and contains 2 fetal arteries and 1 fetal vein.; allowing fetal blood to reach the placenta and exchange molecules with maternal blood.

*Umbilical arteries* transport $CO_2$ and waste molecules to the placenta for disposal.

*Umbilical vein* transports $O_2$ and nutrients from the placenta to the fetal circulatory system.

*Fetal umbilical artery* transports *deoxygenated blood* and carries blood away from fetus's heart.

*Fetal umbilical vein* carries *oxygenated blood* and brings blood to the fetus's heart.

## Allantois

*Allantois* is an extraembryonic membrane that evaginates from the *archenteron* of the gastrula, later fuses with the chorion.

Allantois initially stores waste, such as uric acid, in birds and reptiles.

In mammals, the allantois initially transports waste products to the placenta but regresses in the later stages of gestation. After birth, it may remain a vestigial structure or disappear altogether.

In birds and reptiles, the *allantois* and *chorion* form a double membrane that functions in gas exchange next to the shell.

## Embryonic Gestation

### Early gestation

*Gastrulation* occurs during week 3 of gestation.

*Neurulation* that initiates nervous system development follows gastrulation.

*Neural tube* is visible as a thickening along the entire *dorsal length* of the embryo.

*Ectoderm* develops into *neural folds* as the *neural tube*.

At the end of the first month, the heart starts forming.

Designated cells form the basic structure of the limbs, spine, nervous and circulatory systems.

*Limb buds* appear on arms and legs.

Head enlarges, and the sense organs become more prominent.

Rudiments of the eyes, ears, and nose are evident.

### Gestation weeks four and five

*Heart* development begins and continues into week 4.

Veins enter this tubular heart posteriorly, and the arteries exit anteriorly.

Later the heart twists so that all major vessels are located anteriorly.

Right and left heart tubes fuse, and heart begins pumping blood; chambers are not fully formed.

*Body stalk* is a bridge of mesoderm connecting the *caudal* (or tail) end via the *chorionic villi.*

Head and tail lift; body stalk moves anteriorly by constriction.

*Allantois*, found in the body stalk, extends into the umbilical cord blood vessels.

Organs such as the heart, lungs, and liver continue developing.

### Gestation weeks six to eight

Developing embryos appear more *human-like.*

*Brain* develops, and the head achieves its typical body relationship as the neck region develops.

*Nervous system* exhibits reflex actions like the startle response to touch.

Week 7, the fetus initially exhibits the same external and internal genital structures.

## Müllerian and Wolffian ducts

By week 8, these hormones cause the *Müllerian duct* to degenerate and induce the *Wolffian duct* to develop into the male reproductive tract.

Female fetus with no Y chromosome has gonads that become ovaries; female fetus has external:

> *genital tubercle*

> pair of *urogenital folds*

> pair of *labioscrotal folds*

Week 8, the embryo is about 38 millimeters and weighs no more than an aspirin tablet.

All organs are established.

## Fetal Development

### Gestation weeks ten to twelve

Without anti-Müllerian hormones, the Wolffian ducts degenerate, and Müllerian ducts develop into the female reproductive tract.

*Placenta* is fully formed and produces *progesterone* and *estrogen*.

Negative feedback by hypothalamus and anterior pituitary means *no ovarian follicles mature*.

Instead, *ovarian follicles* maintain *uterus lining* and ensure no menstruation during pregnancy.

*Facial characteristics* of the fetus become recognizable.

### Sexual differentiation during week twelve

> *genital tubercle* differentiates into the *glans* (head) of the *penis* or *clitoris*,
>
> *urogenital folds* enclose the male urethra or become female *labia minora*, and
>
> *labioscrotal folds* become the *scrotum* of a male or *labia majora* of a female.

Sex of the fetus can be identified anatomically.

### XY chromosomal influences

Fetus has an internal:

> pair of *gonadal ridges*,
>
> *mesonephric* or *Wolffian ducts*, and
>
> *paramesonephric* or *Müllerian ducts*.

In XY males, the *SRY gene* encodes for the *testis-determining factor* protein.

*Testis-determining factor protein* initiates the development of male genitalia.

For example, gonadal ridges become a pair of testes, secreting testosterone (male hormone) and *Müllerian-inhibiting factor* (anti-Müllerian hormone).

## Gestation weeks fourteen to sixteen

By week 14, the characteristics of the fetus have mostly developed.

A fetus can only flex its limbs; later, it moves so vigorously that the mother can feel movements beginning in the fourth month.

Fetus soon acquires hair, eyebrows, eyelashes, and nails.

Fine, downy hair *lanugo* covers fetal limbs and trunk but sheds shortly before or after birth.

Skin grows so fast that it wrinkles.

*Vernix caseosa* is a waxy substance that protects the skin from watery amniotic fluid.

After 16 weeks, the *fetal heartbeat* is heard by a stethoscope.

## Gestation weeks twenty-four to thirty-eight

At 24 weeks, a fetus born prematurely survives.

However, the lungs are immature and cannot capture $O_2$ adequately.

Fetus now grows in *size rather than complexity*.

At 38 weeks after fertilization, the fetus is mature enough for live birth.

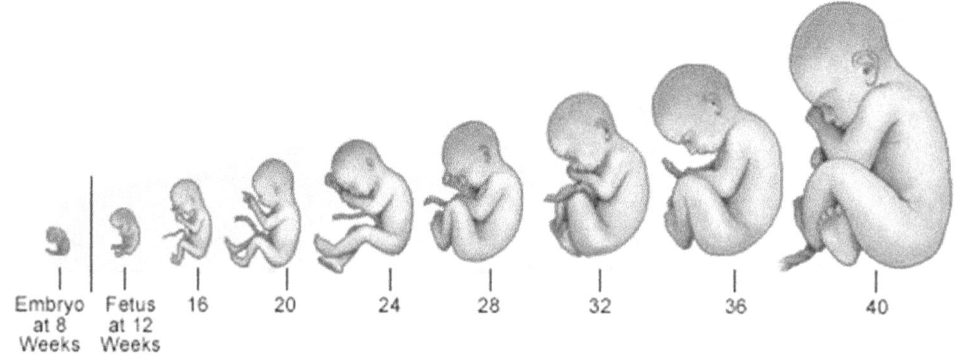


*Fetal growth from week 8 to 40. The gestational period (from conception till birth) in humans is 38 weeks; however, in obstetrics, the term is counted from the first day of the last menstruation (usually 2 weeks earlier), making the pregnancy length 40 weeks.*

## Cellular Communications

### Cell-to-cell communication

*Development of the embryo* is mediated by communication between cells.

*Induction* is an essential mechanism of differentiation using cell-to-cell signaling.

*Inducer* is the cell sending the signal for others to change, while the *responder* (or target) is the cell that receives the signal and changes accordingly.

For example, *optic vesicles* are embryonic brain structures that *induce ectoderm* to develop into eye lenses.

*Induction* may result from *juxtacrine* or *paracrine* mechanisms.

*Juxtacrine* signaling involves communication between two cells that *touch*.

*Paracrine* signaling is when an inducer cell locally releases a substance (*hormone*) that diffuses to a nearby target cell.

### Dorsal lip of blastophore as organizing center

Between 1918-1925, amphibian embryologists Hans Spemann (1869-1941) and his doctoral student Hilde Mangold (1898-1924) experimented on the dorsal side of embryos, where the notochord and the nervous system develop.

*Notochord tissue induces nervous system* formation, even when placed in a different tissue environment, such as beneath the abdominal ectoderm.

Spemann and Mangold showed that the dorsal lip of the blastopore, as the *primary organizer*, is necessary for development.

Spemann was awarded the 1935 Nobel Prize in Physiology or Medicine for discovering the "organizer effect in embryonic development."

*Primary organizer* directs the formation of germ layers during *gastrulation.*

*Germ layers* form in the earliest stages of embryonic development, consisting of *endoderm* (i.e., inner layer), *ectoderm* (i.e., outer layer), and *mesoderm* (i.e., middle layer).

Cells invaginating *first* are closest to the *primary organizer*, so they become *endoderm*.

Cells invaginating *next* (intermediate distance from primary organizer) become *mesoderm*.

Cells furthest from the primary organizer become *ectoderm*.

## Cell polarity and cell migration

For development, it is not enough for cells to differentiate; they must also *migrate.*

*Cell migration* is the controlled movement of cells to predetermined areas.

Migrating cells have polarity and a distinguishable front (*anterior*) and back (*posterior*).

Polarity is necessary for cells to maintain controlled movement.

Without polarity, cells would move simultaneously in various directions.

*Cytoskeletal filaments* establish and maintain a cell's polarity.

Mechanisms of cell migration are not fully understood but foremost are *cytoskeletal* and *membrane flow* models.

## Cytoskeletal model for cell migration

*Cytoskeletal model theory* is based on the action of cytoskeletal elements.

These cytoskeletal elements interact to support and alter a cell's plasma membrane.

Rapid *actin polymerization* of the cytoskeleton at the front edge powers cellular migration.

*Actin filament* (*microfilaments*) formation at the cell front edge is the force for cell movement.

Furthermore, *microtubules* may act to contract the trailing edge of the cell.

Cytoskeletal model theory relies on collaboration by cytoskeletal filaments.

*Microfilaments* push the front edge, and *microtubules* retract the trailing edge.

## Membrane flow theory for cell migration

*Membrane flow theory* uses changes in the plasma membrane rather than the cytoskeleton.

Cells undergo membrane recycling by returning (recycling) the plasma membrane brought into the cell during *endocytosis* to the plasma membrane during *vesicular transport.*

Membrane recycling underlies the membrane flow theory when cell migration occurs by *adding a plasma membrane* to the cell front.

**Integrins and microfilaments**

*Membrane flow model* uses *integrin proteins* attached to the plasma membrane and "walk" the cell along its migratory path.

*Integrins proteins* are principal receptors binding to the extracellular matrix.

Integrins proteins are heterodimers and transmembrane linkers between the *extracellular matrix* and *cytoplasmic actin cytoskeleton*.

Membrane flow model uses *microfilaments* at the cell's front edge as stabilizing agents.

*Integrins* are continually recycled by *endocytosis* to the posterior (rear) as integrins are brought to the cell front by exocytosis.

The two models are not mutually exclusive, so that a hybrid theory may explain cell migration.

*Notes for active learning*

## Stem Cells

### Totipotent and pluripotency stem cells

In early embryogenesis, stem cells are *totipotent.*

*Totipotent cells* can differentiate into any cell type.

Only the *zygote (fertilized egg)* and subsequent *morula (solid ball)* cells are *totipotent.*

Totipotent stem cells can give rise to an entire *indeterminate cleavage* organism.

After a few developmental cell divisions, embryonic stem cells become *pluripotent.*

*Pluripotent stem cells* cannot individually develop into an entire organism but may *differentiate* into any *embryonic cell.*

*Pluripotent stem cells* can become any of the three germ layers.

### Unipotent adult stem cells

*Pluripotent embryonic cells* divide and commit (i.e., determined) to a cell type.

Pluripotent cells become committed to a single cell type and are *unipotent stem cells.*

Unipotent stem cells *self-renew* but give rise to only *one cell type* or *tissue.*

For example, a unipotent skin cell can self-renewal but differentiates to only skin cell fates.

*Unipotent stem cells* represent most human adult stem cells.

### Multipotent adult stem cells

Adults have stem cells; their pluripotency is debated.

Most adult stem cells are *multipotent.*

Multipotent cells can develop into more than one cell type but are more limited than pluripotent.

Adult stem cells and cord blood stem cells are considered multipotent.

*Multipotent stem cells* are between pluripotent and unipotent.

*Multipotent stem cells* (e.g., germ layers) give rise to specific tissue lineages (e.g., dermal layer or lineage within a dermal layer).

Multipotent stem cells have a narrow range of cell types but benefit the organism by regenerating and repairing damaged tissue.

Adult stem cells are in tissues that must be frequently replaced (e.g., skin, bone marrow, liver).

## Medical applications of stem cells

Adult and embryonic stem cells are the subject of intense research into their medical applications for tissue repair and treatment of degenerative diseases.

Stem cells are:

*multipotent* (able to give rise to multiple cells within a lineage),

*pluripotent* (able to give rise to all cell types in an adult), or

*totipotent* (able to give rise to all embryonic and adult lineages).

## Determined and differentiated stem cells

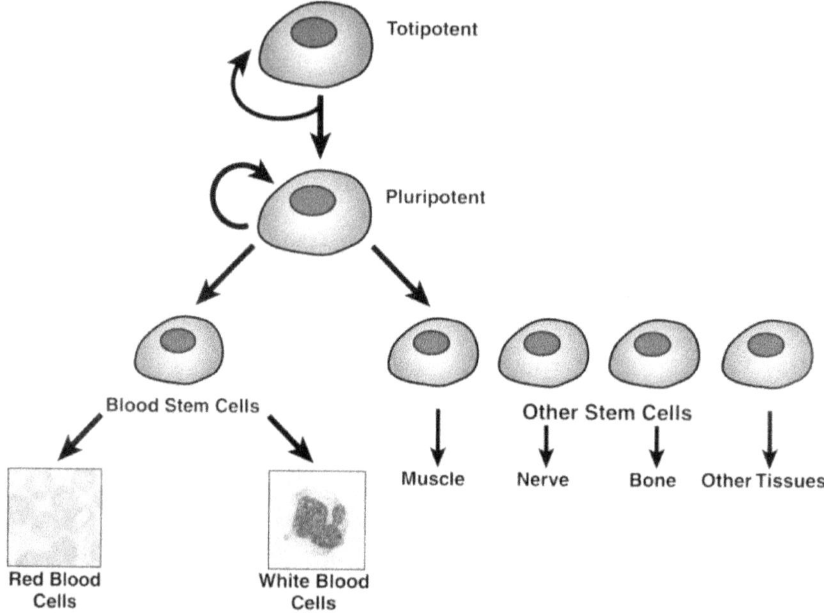


*Stem cells undergo determination before differentiation, where they exhibit the morphology and biochemistry of the specialized (differentiated) cell*

## Gene Regulation

### Gene regulation in development

*Gene regulation* of embryonic cells drives cellular migration and differentiation.

Developmental gene regulation has been based on *Caenorhabditis elegans* (*roundworm*) research by 2002 Nobel laureate Sydney Brenner (1927-2019) and *Drosophila melanogaster* (*fruit fly*) by 1933 Nobel laureate Thomas Hunt Morgan (1866-1945).

*Maternal determinants* are usually homogeneously distributed throughout the ovum.

However, *Drosophila melanogaster* exhibits cytoplasmic segregation of maternal determinants.

*Drosophila* ovum (oocyte) asymmetrically distributes *maternal effect genes.*

### Maternal effect genes

*Maternal effect genes* produce mRNA and proteins (maternal determinants) in the mature ovum, remaining asymmetrically distributed.

Maternal determinants influence the expression of specific zygotic genes; the mother's genotype influences the zygote's phenotype.

Maternal effect genes regulate the first class of zygotic genes, *gap genes.*

*Gap gene* proteins regulate *pair-rule genes*, influencing transcription of *segment polarity genes.*

### Gap, pair-rule, and segment polarity genes

*Gap genes, pair-rule genes,* and *segment polarity genes* interact to influence *homeotic genes.*

*Gap, pair-rule, segment polarity genes,* and *homeotic genes* partition the developing *Drosophila* embryo into body segments (i.e., *pattern formation*).

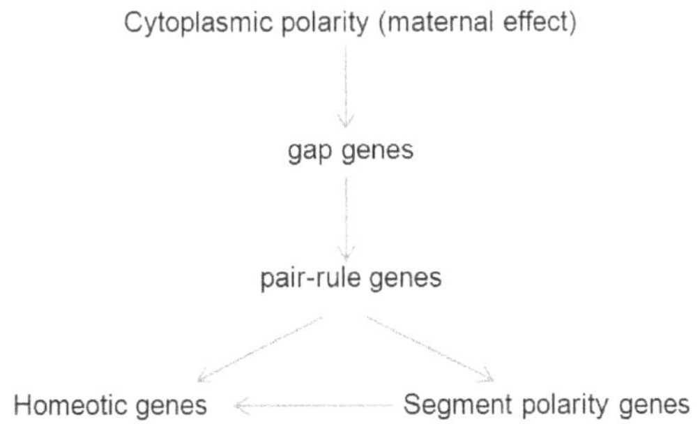


*Gene regulation in development uses upstream gene expression and protein production*

Function has been studied by mutating genes and observing body plan malfunctions.

Maternal determinants' initial gradients are not the only gene regulation in developing embryos.

Gene regulation is a complex process due to how genes may be regulated.

## Gene Expression

### Histone modifications regulate gene expression

During DNA to protein, and even after protein synthesis, gene expression can be regulated.

*Histone modifications* are the highest level of transcriptional regulation.

Histones are positively charged DNA-binding proteins.

Histones make genes on DNA accessible (or inaccessible) to RNA polymerase.

Histones modulate transcription factor binding to DNA and *facilitate* (or *inhibit*) RNA polymerase binding to *initiate (or impede) transcription.*

### Post-transcriptional gene regulation

*Post-transcriptional regulation* occurs after transcription and before translation, targeting *mRNAs transcript* moving from the nucleus to cytoplasm.

Post-transcriptional regulation may be accomplished by:

> *nuclear export* modulation (though nuclear envelope pores)
>
> RNA *splicing* (joining exons and removing introns)
>
> RNA *editing* (alternative spicing of primary transcripts)
>
> capping and polyadenylation (protecting RNA from degradation in the cytoplasm)

mRNA is processed (splicing of exons and removing introns, adding a 5' G cap and 3' poly–A tail) and exported from the nucleus.

Cell may alter gene regulation at the *translational level* by promoting (or inhibiting) *ribosome recruitment* to the mRNA.

Ribosomes convert nucleotides (genetic DNA and mRNA) to amino acids (proteins).

### Post-translational gene regulation

Gene regulation does not cease once a protein is translated.

Post-translational gene regulation acts upon proteins that have been synthesized.

For example, proteins may be sequestered in cellular regions or assembled with proteins.

Additionally, *zymogens* are enzymes cleaved (or modified) to become active.

Post-translational modifications target *histones* (negatively charged DNA-chromatin proteins).

**Histone phosphorylation and acetylation**

Histone modifications include:

*phosphorylation* (adding negatively charged phosphates) and

*acetylation*, which drives chromatin rearrangement.

Histone modifications are an active area of current scientific research.

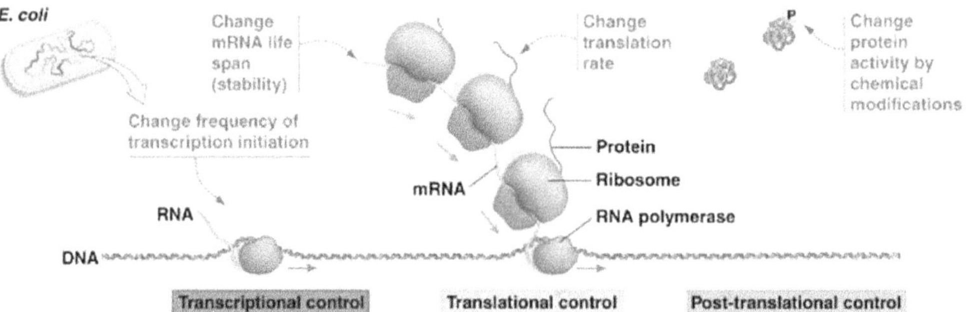


*Regulation of gene expression: transcriptional (DNA → RNA), post-transcriptional (e.g., splicing), post-translational control (e.g., glycosylation).*

## Environmental Influences in Development

### Environment–gene interactions

Genetic and environmental factors contribute to the complex etiology of congenital disabilities by disrupting the highly regulated embryonic developmental processes.

Intrauterine environment of the developing embryo and fetus is subject to maternal health status, lifestyle, medications, maternal genotype, and exposure to environmental teratogens.

*Teratogens* are agents which cause *congenital abnormalities* (i.e., present since birth).

Teratogens may cause mild to severe congenital disabilities or terminate the pregnancy.

For example, radiation, chemicals, certain drugs, and pathogens may be classified as teratogens.

*Congenital disabilities* (1 in 33 babies in the US) continue to be the leading cause of infant death (about 5,500 per year) in the U.S. and a field of intense biomedical and clinical research.

Studies suggest that a relationship between genetics and environmental factors propagates congenital disabilities.

### Congenital disabilities

*Teratology* focuses on the causes and underlying mechanisms of congenital disabilities for decades, yet understanding these critical issues remains vague.

Organs have a sensitive period when toxic substances alter their development.

Placenta is an effective mediator between the fetus and mother but is highly susceptible to chemical penetration.

Chemical influences are of concern during the *embryonic period* (fertilization through 10th week of gestation; or the 8th week of embryonic age).

*Embryonic period* is when crucial structures are in vulnerable stages of development.

For example, between 1958 and 1961, it was typical for pregnant women in England and other countries to take the tranquilizer thalidomide between days 27 and 40 of pregnancy.

Thalidomide is a potent teratogen, and infants are born with deformed limbs.

After this period, the fetus is resistant to thalidomide effects, and the child is born unaffected.

Studying gene-environment interactions elucidates the complex biological mechanisms and pathological processes contributing to congenital disabilities.

*Notes for active learning*

## Aging and Regeneration

### Programmed cell death

*Apoptosis* is *programmed cell death* and a part of development.

Apoptosis is central to gametogenesis, embryogenesis, removal of damaged cells, and aging.

During apoptosis, *caspases* (or *proteases*) and *nucleases* digest the cell from within.

These protease enzymes are inactive in a typical cell.

Proteases are activated by a complex *signal transduction pathway*, which may be initiated by external commands or by internal recognition of cell malfunction, infection, or DNA damage.

Apoptosis in animal embryogenesis and fetal development has been extensively studied.

For example, fate maps of *C. elegans* show 131 cells undergo apoptosis for worm development.

Cell-death protein is normally inhibited, but when the cells are signaled via induction, the cell-death cascade becomes active and produces proteases and nucleases to slice proteins and DNA.

### Cellular changes during apoptosis

Apoptotic cells undergo characteristic changes:

> *protrusion of the plasma membrane* (*blebbing*)
>
> *shrinkage*
>
> *nuclear fragmentation*
>
> *chromatin condensation*
>
> *chromosomal DNA fragmentation*

Apoptotic cells attract nearby cells, which consume the dying cell to prevent harmful cell contents from spilling into the extracellular fluid and affecting other cells.

## Apoptosis for tissue development

During embryonic development, apoptosis removes harmful, abnormal, or unneeded cells.

Dysregulation of apoptosis during gestation can lead to severe congenital disabilities.

Apoptosis begins development when cells in the blastula die off to form the gastrula.

Morphogenesis uses apoptosis; organs overproduce cells and "prune" excess cells to sculpt the intricate shape and function of organs.

For example, fingers and toes are webbed early in gestation but lose their webbing and become recognizable digits.

Another example of apoptosis-assisted morphogenesis occurs in the center of ducts and tubes to hollow them out.

Many fetal structures must be shed via apoptosis before birth.

For example, female fetuses degrade Wolffian ducts, while males degrade Mullerian ducts.

## Apoptosis destroys malignant cells

Regulated apoptosis prevents malignant cells from dividing and invading other tissues, leading to malignant tumors.

Apoptosis continues in adults for tissue remodeling and immune system maintenance.

For example, cancer, autoimmune diseases, and neurodegenerative diseases result from dysregulated apoptosis.

Excessive apoptosis leads to tissue degeneration; malfunctions implicated in aging and disease.

## Regenerative capacity in species

*Regeneration* is the reactivation of development to regrow damaged body structures.

While regenerative abilities are limited in humans, many animals retain the capacity to regrow entire limbs after embryonic development.

For example, most echinoderms have robust regenerative powers. Some sea stars can regrow an entirely new organism from a single arm, while others can only replace lost limbs.

Separated limbs must survive nutrient stores until they regenerate. Regrowth is not assured and takes years. Sea stars are vulnerable to infection during regrowth.

Newts and salamanders (particularly the axolotl) undergo regeneration in two stages.

*First*, adult cells at the site of the limb's separation de-differentiate into *progenitor cells*.

*Second*, the progenitor cells proliferate and differentiate until they replace the missing tissue.

## Human regeneration

Humans exhibit regeneration via *progenitor cells*, although regrowth of limbs has not occurred.

For example, the liver has remarkable regeneration capacities. The liver restores lost mass through complicated signaling cascades and adjusts size while fulfilling its duties.

Human embryos are *totipotent* and can regenerate complete organs and limbs.

However, like many vertebrates, this ability is lost during embryogenesis, leaving adult humans with a limited capacity to restore tissue damage.

## Cellular mortality

*Senescence* is the gradual deterioration of function in an organism, eventually leading to death.

Except for a few with remarkable properties of immortality, all organisms undergo senescence.

One puzzling question is *why* people age; reasons for senescence are likely combined factors.

Some research suggests that senescence avoids cancer, while others cite environmental factors, such as radiation and oxidative agents, causing DNA and cellular damage.

Senescence refers to a cell that can *no longer divide* at the cellular level.

A senescent cell is not dead; it actively continues its metabolic functions.

Senescent may initiate by activating oncogenes or the cell's recognition of DNA damage.

*Oncogenes* signal abnormal cells to avoid senescence or undergo *apoptosis to avoid tumors*.

## Telomerase and Hayflick limit

DNA damage causes include radiation, oxidation (free radicals), and telomeres shortening.

*Telomeres* are nonsensical DNA of 300 sequence repeats added to DNA 3' ends by *telomerase*.

Replication shortens DNA, leading to severe DNA damage so the cell cannot continue dividing.

Telomeres prevent DNA loss from replication to protect the encoding regions from destruction.

Telomerase allows cells to proliferate because DNA replication gene loss does not affect them.

*Hayflick limit* states that cells are limited to 50 to 70 mitotic divisions.

Cells are subject to the *Hayflick limit*, but embryonic cells and specific adult cells contain high telomerase levels to divide continually.

Telomere shortening is associated with diseases involving premature aging, such as pulmonary fibrosis (scarring of the lungs).

**Cellular senescence**

Shortened telomeres cause cells to divide with gene loss, leading to cellular senescence.

Lack of sufficient telomeres causes chromosomes to fuse, corrupt genetics, and appear broken.

Cells recognize DNA damage and thus enter apoptosis.

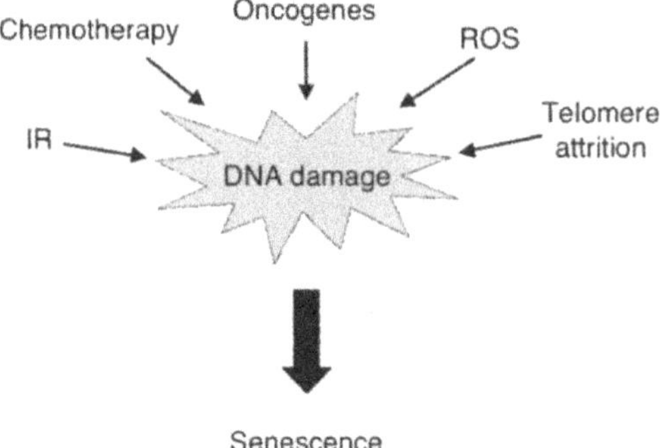



*Environmental and biochemical factors contributing to senescence*

Jellyfish, flatworms, and hydrae are *biologically immortal*, exhibiting negligible senescence, and do not age.

*Biologically immortal* organisms cannot escape death because they experience disease, injury, and predators.

Intense research studies model therapies based on these organisms' *"immortal"* cells to address aging and mortality.

For the present, however, senescence is inevitable with human development.

*Notes for active learning*

*Notes for active learning*

# PRACTICE QUESTIONS

# &

# DETAILED EXPLANATIONS

## Practice Questions: Reproduction

1. Which event is essential for the menstrual cycle?

> I.   adrenal medulla releases norepinephrine
> II.  FSH stimulates ovarian follicle development
> III. progesterone stimulates the formation of the endometrial lining

A. II only
B. III only
C. I and II only
D. II and III only
E. I and III only

2. All the following about human gamete production is true, EXCEPT:

A. meiosis in females produces four egg cells
B. sperm develops in the seminiferous tubules within the testes
C. eggs develop in the ovarian follicles within the ovaries
D. FSH stimulates gamete production in both males and females
E. gametes arise via meiosis

3. All these statements regarding the menstrual cycle are true, EXCEPT:

A. Graafian follicle, under the influence of LH, undergoes ovulation
B. follicle secretes estrogen as it develops
C. corpus luteum develops from the remains of the post-ovulatory Graafian follicle
D. FSH causes the development of the primary follicle
E. FSH and LH are both secreted by the posterior pituitary

4. Which is the initial site of spermatogenesis?

A. seminiferous tubules
B. seminal vesicles
C. vas deferens
D. epididymis
E. prostate

5. Which event occurs first?

A. formation of corpus luteum
B. rupture of the Graafian follicle
C. secretion of estrogen
D. release of progesterone
E. decreased FSH release by the pituitary

**6.** Which statement correctly describes the role of LH in the menstrual cycle?

    **A.** stimulates the ovary to increase LH secretions

    **B.** inhibits secretions of GnRH

    **C.** stimulates the development of the endometrium for implantation of the zygote

    **D.** induces the corpus luteum to secrete estrogen and progesterone

    **E.** stimulates milk production after birth

**7.** What causes the testes to not fully descend into the scrotum due to abnormal testicular development?

    **A.** cortisol deficiency

    **B.** testosterone deficiency

    **C.** excess LH

    **D.** excess estrogen

    **E.** excess FSH

**8.** Testosterone is synthesized primarily by:

    **A.** sperm cells

    **B.** hypothalamus

    **C.** Leydig cells

    **D.** anterior pituitary gland

    **E.** seminiferous tubules

**9.** Which condition is the LEAST probable cause of male infertility?

    **A.** acrosomal enzymes denaturation

    **B.** immotility of cilia

    **C.** abnormal mitochondria

    **D.** testosterone deficiency

    **E.** abnormal flagellum

**10.** The surgical removal of the seminiferous tubules would likely cause:

    **A.** sterility, because sperm would not be produced

    **B.** sterility, because sperm would not be able to exit the body

    **C.** reduced volume of semen

    **D.** enhanced fertilization potency of sperm

    **E.** testes to migrate back into the abdominal cavity

**11.** Which results from scars in women's reproductive system with an elevated risk of infections (e.g., chlamydia)?

    **A.** elevated levels of estrogen

    **B.** decreased ovulation

    **C.** infertility

    **D.** reduced gamete production

    **E.** decreased levels of estrogen

**12.** The surgical removal of the seminal vesicles would likely cause:

    **A.** sterility, because sperm would not be produced

    **B.** sterility, because sperm would not be able to exit the body

    **C.** testes to migrate back into the abdominal cavity

    **D.** enhanced fertilization potency of sperm

    **E.** reduced volume of semen

**13.** The primary difference between estrous and menstrual cycles is that:

    **A.** in estrous cycle, the endometrium is shed and reabsorbed by the uterus, while in the menstrual cycle the shed the endometrium is excreted from the body

    **B.** behavioral changes during estrous cycles are much less apparent than those during menstrual cycles

    **C.** season and climate have less pronounced effects on the estrous cycle than they do on menstrual cycles

    **D.** copulation typically occurs across the estrous cycle, whereas in menstrual cycles, copulation only occurs during the period surrounding ovulation

    **E.** most estrous cycles are much longer in duration compared to menstrual cycles

**14.** One function of estrogen is to:

    **A.** induce the ruptured follicle to develop into the corpus luteum

    **B.** stimulate testosterone synthesis in males

    **C.** maintain female secondary sex characteristics

    **D.** promote the development and release of the follicle

    **E.** lower blood glucose

**15.** Increasing estrogen levels during the female menstrual cycle trigger which feedback mechanism?

    **A.** positive, which stimulates LH secretion by the anterior pituitary

    **B.** positive, which stimulates FSH secretion by the anterior pituitary

    **C.** negative, which stimulates the uterine lining to be shed

    **D.** negative, which inhibits progesterone secretion by the anterior pituitary

    **E.** none of the above

**16.** The epididymis functions to:

**A.** synthesize and release testosterone

**B.** store sperm until they are released during ejaculation

**C.** initiate the menstrual cycle by secreting FSH and LH

**D.** provide a conduit for the ovum as it moves from the ovary into the uterus

**E.** provide most of the fluid that comprises semen

**17.** What is the purpose of the cilia covering the inner linings of the Fallopian tubes?

**A.** Preventing polyspermy by immobilizing additional incoming sperm after fusion

**B.** Facilitating movement of the ovum toward the uterus

**C.** Removing particulate matter that becomes trapped in the mucus layer

**D.** Protecting the ovum from pH fluctuations

**E.** Protecting the ovum from temperature fluctuations

**18.** Which endocrine structure initiates the production of testosterone?

**A.** adrenal medulla

**B.** hypothalamus

**C.** pancreas

**D.** anterior pituitary

**E.** posterior pituitary

**19.** Decreasing progesterone during the luteal phase of the menstrual cycle results from:

**A.** increased secretion of estrogen in the follicle followed by the menstruation phase

**B.** degeneration of the corpus luteum in the ovary

**C.** increased secretion of LH, which produces the luteal surge and onset of ovulation

**D.** thickening of the endometrial lining in preparation for implantation of the zygote

**E.** none of the above

**20.** Vasectomy prevents the following:

**A.** movement of sperm along the vas deferens

**B.** transmission of sexually transmitted disease

**C.** production of sperm via spermatogenesis

**D.** production of semen in seminal vesicles

**E.** synthesis of seminal fluid

**21.** How many double-stranded DNA molecules are in a single mouse chromosome after gametes form?

A. 0

B. 1

C. 2

D. 4

E. 8

**22.** Common red-green color blindness is an X-linked trait. When a woman whose father is color blind has a son with a non-afflicted man, what is the probability that their son will be color blind?

A. 0

B. 1/4

C. 1/2

D. 3/4

E. 1

**23.** The likely gamete cell to be produced from meiosis in the seminiferous tubules is:

A. diploid 2° spermatocytes

B. haploid 1° spermatocytes

C. haploid spermatids

D. diploid spermatids

E. none of the above

**24.** At birth, a woman possesses a finite number of ova. In oogenesis, the meiotic division is arrested at which stage until she reaches menarche?

A. ovum

B. oogonium

C. ova

D. secondary oocytes

E. primary oocytes

**25.** A human cell after the first meiotic division is:

A. 2N and 2 chromatids

B. 2N and 4 chromatids

C. 1N and 2 chromatids

D. 1N and 1 chromatid

E. None of the above

**26.** What distinguishes meiosis from mitosis?

    I. Genetic recombination

    II. Failure to synthesize DNA between successive cell divisions

    III. Separation of homologous chromosomes into distinct cells

**A.** I only

**B.** II only

**C.** I and III only

**D.** II and III only

**E.** I, II and III

**27.** Oocytes within primordial follicles of the ovary are arrested in:

**A.** interphase

**B.** prophase II of meiosis

**C.** prophase I of meiosis

**D.** prophase of mitosis

**E.** G1

**28.** SRY gene encoding for the testis-determining factor is the primary sex-determining gene that resides on:

**A.** pseudoautosomal region of the Y chromosome

**B.** short arm of the Y chromosome, but not in the pseudoautosomal region

**C.** X chromosome

**D.** pseudoautosomal region of the X chromosome

**E.** autosomes

**29.** The number of chromosomes contained in the human primary spermatocyte is:

**A.** 23

**B.** 23, X/23, Y

**C.** 92

**D.** 184

**E.** 46

**30.** Which process does NOT contribute to genetic variation?

**A.** Random segregation of homologous chromosomes during meiosis

**B.** Random segregation of chromatids during mitosis

**C.** Recombination

**D.** Mutation

**E.** All the above

**31.** In human females, secondary oocytes do not complete meiosis II until:

**A.** menarche

**B.** menstruation

**C.** puberty

**D.** menopause

**E.** fertilization

**32.** 47, XXY is a condition known as:

**A.** Turner syndrome

**B.** double Y syndrome

**C.** trisomy X syndrome

**D.** Klinefelter syndrome

**E.** fragile X syndrome

**33.** All cells contain diploid (2N) numbers of chromosomes EXCEPT:

**A.** primary oocyte

**B.** spermatogonium

**C.** spermatid

**D.** zygote

**E.** oogonium

**34.** The probability that all children in a four-children family will be males is:

**A.** 1/2

**B.** 1/4

**C.** 1/8

**D.** 1/16

**E.** 1/64

**35.** Unequal cytoplasm division is characteristic of:

**A.** binary fission of bacteria

**B.** mitosis of a kidney cell

**C.** production of sperm

**D.** production of an ovum

**E.** none of the above

**36.** All are clinical manifestations of Kartagener's syndrome, resulting from defective dynein that causes paralysis of cilia and flagella, EXCEPT:

**A.** chronic respiratory disorders

**B.** cessation of ovulation

**C.** male infertility

**D.** ectopic pregnancy

**E.** middle ear infections

**37.** Translation, transcription, and replication occur in which phase of the cell cycle?

A. G1

B. G2

C. metaphase

D. anaphase

E. S

**38.** Progesterone is primarily secreted by the:

A. primary oocyte

B. hypothalamus

C. corpus luteum

D. anterior pituitary gland

E. endometrial lining

**39.** All statements are true for a normal human gamete, EXCEPT it:

A. originates via meiosis from a somatic cell

B. contains genetic material that has undergone recombination

C. contains a haploid number of genes

D. always contains an X or Y chromosome

E. forms from the meiotic process

**40.** Which cell division results in four genetically different daughter cells with 1N chromosomes?

A. interphase

B. somatic cell regeneration

C. cell division

D. mitosis

E. meiosis

**41.** A genetically important event of crossing over occurs during:

A. metaphase II

B. telophase I

C. anaphase I

D. prophase I

E. telophase II

**42.** Which does NOT describe events occurring in prophase I of meiosis?

A. chromosomal migration

B. genetic recombination

C. formation of a chiasma

D. spindle apparatus formation

E. tetrad formation

**43.** The difference between spermatogenesis and oogenesis is:

**A.** spermatogenesis produces haploid cells, while oogenesis produces diploid cells
**B.** spermatogenesis produces gametes, while oogenesis does not produce gametes
**C.** oogenesis is a mitotic process, while spermatogenesis is a meiotic process
**D.** spermatogenesis is a mitotic process, while oogenesis is a meiotic process
**E.** spermatogenesis produces 4 1N sperms, while oogenesis produces 1 egg and polar bodies

**44.** How many Barr bodies are present in the white blood cells of a 48, XXYY individual?

**A.** 0
**B.** 1
**C.** 2
**D.** 3
**E.** 4

**45.** Polar bodies are the products of:

**A.** meiosis in females
**B.** meiosis in males
**C.** mitosis in females
**D.** mitosis in males
**E.** two of the above

**46.** Turner syndrome results from:

**A.** extra chromosome number 13
**B.** absence of the Y chromosome
**C.** presence of an extra Y chromosome
**D.** trisomy of the X chromosome
**E.** monosomy of an X chromosome

**47.** A sheep in 1996 named Dolly was different from animals produced by sexual reproduction because:

**A.** her DNA is identical to the DNA of her offspring
**B.** she was carried by a surrogate mother, while her DNA came from two other individuals
**C.** all her cells' DNA are identical
**D.** her DNA was taken from a somatic cell of an adult individual
**E.** her DNA was taken from a gamete of a neonatal individual

*Notes for active learning*

## Detailed Explanations: Reproduction

**Answer Keys**

| | | | | |
|---|---|---|---|---|
| 1:  D | 11: C | 21: B | 31: E | 41: D |
| 2:  A | 12: E | 22: C | 32: D | 42: A |
| 3:  E | 13: A | 23: C | 33: C | 43: E |
| 4:  A | 14: C | 24: E | 34: D | 44: B |
| 5:  C | 15: A | 25: C | 35: D | 45: A |
| 6:  D | 16: B | 26: D | 36: B | 46: E |
| 7:  B | 17: B | 27: C | 37: E | 47: D |
| 8:  C | 18: B | 28: B | 38: C | |
| 9:  B | 19: B | 29: E | 39: A | |
| 10: A | 20: A | 30: B | 40: E | |

### 1.  D is correct.

*Menstrual cycle* is regulated by the hypothalamus, anterior pituitary, and ovaries.

*Follicle-stimulating hormone* (FSH) is released by the anterior pituitary, stimulating the maturation of an ova.

*Progesterone*, synthesized by the ovary and corpus luteum, regulates the development and shedding of the endometrial lining of the uterus.

*Adrenal medulla* synthesizes epinephrine (adrenaline) and norepinephrine, which assist the sympathetic nervous system in stimulating the *fight or flight* response but are *not* involved in the menstrual cycle.

### 2.  A is correct.

*Oogenesis* produces one viable egg and (up to) three polar bodies resulting from the cytoplasm's unequal distribution during meiosis.

*Gametes* (e.g., egg and sperm) become haploid (1N) through reductive division (i.e., meiosis), in which a diploid cell (2N) gives rise to four haploid sperm or one haploid egg and (up to) three polar bodies.

B: *interstitial cells* are stimulated by luteinizing hormone (LH) to produce testosterone, stimulating sperm development within the seminiferous tubules, along with FSH.

C: *follicle-stimulating hormone* (FSH) stimulates eggs to develop in follicles within ovaries.

D: *follicle-stimulating hormone* (FSH) participates in gamete production for males and females.

### 3.  E is correct.

*Anterior pituitary* secretes *follicle-stimulating hormone* (FSH) and *luteinizing hormone* (LH) to influence the maturation of the follicle.

*Follicle-stimulating hormone* (FSH) stimulates the maturation of gametes (e.g., ova and spermatids).

Follicle-stimulating hormone (FSH) stimulates estrogen production, which aids in maturing the primary follicle.

*Luteinizing* (LH) stimulates ovulation and the development of the corpus luteum.

*Mature corpus luteum* secretes progesterone, causing the uterine lining to thicken and become vascular in preparation for the implantation of the fertilized egg (i.e., zygote).

### 4.  A is correct.

*Sperm synthesis* begins within the seminiferous tubules of the testes.

Sperm undergoes maturation and storage in the *epididymis*.

During ejaculation, sperm are released from the *vas deferens*.

*Seminal vesicles* are a pair of male accessory glands that produce about 50-60% of the liquid of the semen.

Mnemonic SEVEN UP — sperm path during ejaculation:

> **S**eminiferous tubules > **E**pididymis > **V**as deferens > **E**jaculatory duct > **N**othing > **U**rethra > **P**enis

### 5.  C is correct.

*Hypothalamus* secretes releasing factors (tropic hormones), so the anterior pituitary releases LH and FSH.

*Follicle-stimulating hormone* (FSH) stimulates ovaries to produce mature *ovarian follicles*.

During the follicular stage, the ovary produces estrogen.

As estrogen is released, FSH levels drop, and LH levels increase, which triggers the follicle to release the ovum (i.e., ovulation).

*Luteinizing hormone* (LH) affects the corpus luteum (i.e., formerly follicle), secreting progesterone.

*Estrogen* is first secreted by ovaries (under the influence of FSH) during the follicular stage (days 1 to 14), decreasing FSH secretions.

*continued...*

*Decrease in FSH*, along with an increase in LH, causes rupture of the follicle (i.e., ovulation) and formation of the *corpus luteum*.

As the corpus luteum matures, *progesterone levels increase.*

## 6.  D is correct.

At day 14, LH surges, causing a mature follicle to burst and release the ovum from the ovary (i.e., ovulation).

Following ovulation, luteinizing hormone (LH) induces the ruptured follicle to develop into the corpus luteum, secreting progesterone and estrogen.

*Prolactin* stimulates milk production after birth.

*Ovaries secrete estrogen and progesterone*, while the *anterior pituitary secretes LH*.

Progesterone and estrogen *inhibit GnRH release*, inhibiting the release of FSH and LH and preventing additional follicles from maturing.

*Progesterone* stimulates the development and maintenance of endometrium in preparation for embryo implantation.

## 7.  B is correct.

*Luteinizing hormone* (LH) stimulates the release of testosterone in males by Leydig cells.

*Testosterone* is necessary for proper development of the testes, penis, and seminal vesicles.

Testosterone surges between the first and fourth months of life, and testosterone inadequacy is a primary cause of *cryptorchidism.*

During puberty, testosterone is needed for *secondary male sex characteristics* (e.g., growth of body hair, broadening of shoulders, enlarging of the larynx, deepening voice, and increased secretions of oil and sweat glands)

*Cortisol* is a stress hormone that elevates blood glucose levels.

E: FSH stimulates the maturation of the gametes (e.g., ova and spermatids).

## 8.  C is correct.

*Leydig cells* are *androgens* (i.e., male hormones such as 19-carbon steroids) that secrete testosterone, androstenedione, and dehydroepiandrosterone (DHEA) when stimulated by luteinizing hormone (LH) released by the pituitary gland.

Leydig cells are adjacent to *seminiferous tubules* in testicles.

*continued...*

*Luteinizing hormone* (LH) increases conversion of cholesterol to pregnenolone leading to testosterone synthesis and secretion by Leydig cells.

*Prolactin* (PRL) increases the response of Leydig cells to LH by increasing the number of LH receptors expressed on Leydig cells.

## 9. B is correct.

*Spermatozoon* development produces motile, mature sperm, fusing with ovum to form zygote.

Eukaryotic *cilia* are structurally like eukaryotic flagella, but distinctions are based on function and length.

*Microtubules* form the sperm's flagella, necessary for movement through the cervix, uterus, and along the fallopian tubes (i.e., oviducts).

*Acrosomal enzymes* digest the outer zona pellucida (a glycoprotein membrane surrounding the plasma membrane of an oocyte) to permit the fusion of sperm and egg.

*Sperm midpiece* (diagram below) has many mitochondria used for ATP production for the sperm's movement (via the flagellum) through the female cervix, uterus, and Fallopian tubes.

D: *testosterone* is necessary for *spermatogenesis*.

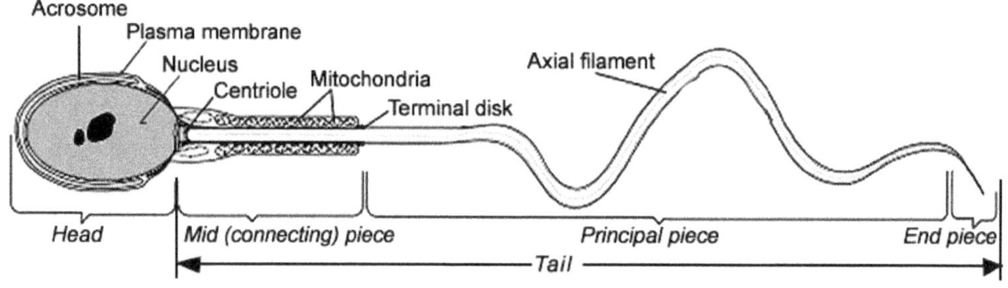


*Human spermatozoon*

## 10. A is correct.

*Seminiferous tubules* are in the testes where meiosis occurs to create gametes (e.g., spermatozoa).

Epithelium of *seminiferous tubules* consists of *Sertoli* (i.e., *nurse cells*) cells that nourish developing sperm and function as phagocytes by consuming residual cytoplasm during spermatogenesis.

Between the Sertoli cells are spermatogenic cells, which differentiate through meiosis into sperm cells.

## 11. C is correct.

*Chlamydia* and other infections may scar the reproductive tract, preventing the ova from reaching the uterus and resulting in infertility.

## 12. E is correct.

*Seminal vesicles* secrete a considerable proportion of the fluid that ultimately becomes semen.

50-70% of the seminal fluid originates from the seminal vesicles but is not expelled in the first ejaculate fractions dominated by spermatozoa and zinc-rich prostatic fluid.

## 13. A is correct.

*Estrous cycle* (i.e., sexual desire) comprises the recurring physiologic changes induced by reproductive hormones in most mammalian females.

Estrous cycles start after sexual maturity in females and are interrupted by anestrous phases or pregnancies.

*Menstrual cycle* (diagram below) changes the uterus and ovary for sexual reproduction.

The menstrual cycle is essential for producing eggs and preparing the uterus for pregnancy.

In humans, the length of a menstrual cycle varies significantly among women (ranging from 21 to 35 days), with 28 days designated as the average length.

Each cycle has three phases based on the ovary (ovarian cycle) or the uterus (uterine cycle).

*Ovarian cycle* consists of the *follicular, ovulation, and luteal phases*, whereas the uterine cycle has *menstruation, proliferative,* and *secretory phases*.

Endocrine system controls both cycles, and the regular hormonal changes can be interfered with using hormonal contraception to prevent reproduction.

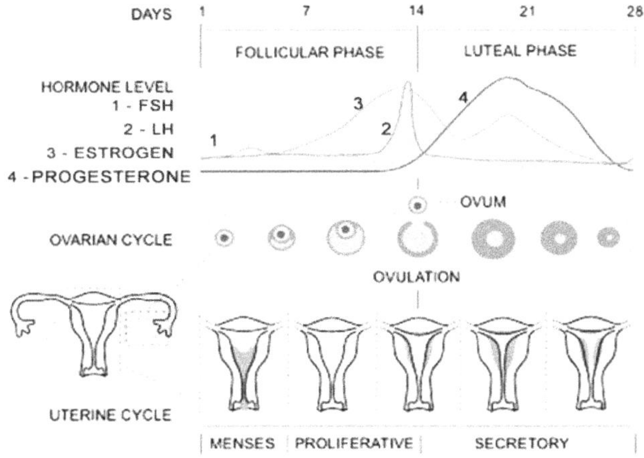


*Menstrual cycle with follicular and luteal phases*

**14.  C is correct.**

*Estrogen* is a steroid hormone necessary for typical female maturation.

Estrogen stimulates the development of the female reproductive tract and contributes to *secondary sexual characteristics* and *libido*.

Estrogen is secreted by the *follicle* during the menstrual cycle and is responsible for the thickening of the endometrium in preparation for the implantation of the fertilized egg.

A: *luteinizing hormone* (LH) induces the ruptured follicle to develop into the *corpus luteum*.

B: *luteinizing hormone* stimulates testosterone synthesis in males.

D: *follicle-stimulating hormone* (FSH) is released by the anterior pituitary and promotes the development of the follicle, which matures and begins secreting estrogen.

E: *insulin* is a peptide hormone secreted in response to high blood glucose levels.

Insulin stimulates the uptake of glucose by muscle and adipose cells and the conversion of glucose into the storage molecule of glycogen.

**15.  A is correct.**

Increased estrogen secretions initiate the luteal surge that increases LH secretion → ovulation.

**16.  B is correct.**

Sperm mature and are stored until ejaculation in the *epididymis*.  During ejaculation, sperm flows from the lower portion of the epididymis. Products from the prostate gland have not activated them, and they cannot swim.

Sperm is *transported via peristaltic action* of muscle layers within the *vas deferens* and forms *semen* mixed with diluting fluids of the seminal vesicles and other accessory glands before ejaculation.

*Seminiferous tubules* secrete testosterone.

Primary functions of the testes are to produce sperm (i.e., spermatogenesis) and androgens (e.g., testosterone).

*Leydig cells* are localized between seminiferous tubules and produce and secrete testosterone and other androgens (e.g., DHT, DHEA) necessary for sexual development and puberty, secondary sexual characteristics (e.g., facial hair, sexual behavior, and libido), supporting spermatogenesis and erectile function.

Luteinizing hormone (LH) results in testosterone release.

Testosterone and follicle-stimulating hormone (FSH) are needed to support spermatogenesis.

**17. B is correct.**

*Inner lining of the Fallopian tubes* is covered with cilia, helping the egg move toward the uterus, where it implants if fertilized.

*Fertilization* usually happens when sperm fuses with an egg in the Fallopian tube.

C: cilia lining the respiratory tract perform this function.

D: *Fallopian tubes* are isolated from the external environment, so pH fluctuations are not an issue.

**18. B is correct.**

In males, *testosterone* is primarily synthesized in *Leydig cells* in the testis.

Testosterone synthesis is regulated by the *hypothalamic–pituitary–testicular axis.*

When testosterone levels are low, the hypothalamus releases *gonadotropin-releasing hormone* (GnRH), which stimulates the pituitary gland to release FSH and LH.

FSH and LH stimulate the testis to produce testosterone.

Testosterone acts on the hypothalamus and pituitary with elevated levels inhibiting GnRH and FSH/LH release through a negative feedback loop.

**19. B is correct.**

If fertilization of the ovum does not occur, the *corpus luteum stops secreting progesterone and degenerates.*

*Menstruation phase* follows decreased progesterone secretion but does not result from increased estrogen levels.

Increased estrogen secretion (not LH) causes luteal surge, which occurs earlier in the cycle.

Thickening of the endometrial lining occurs while estrogen and progesterone levels are high.

**20. A is correct.**

*Vasectomy* procedures prevent the movement of sperm along the vas deferens.

**21. B is correct.**

*Gametes* form via *meiosis* and are double-stranded haploids.

A single chromosome consists of two hydrogen-bonded complementary antiparallel DNA strands.

## 22. C is correct.

Let C designate wild-type and c designate the color-blind allele.

Mother is Cc, and the father is CY (a normal allele with a single copy of the X gene).

From mating, the mother's gamete (as a carrier due to her dad) is C or c, with a 50% probability of the gamete inheriting the C or c allele.

Assuming a boy (i.e., the father transmits Y and not X), the father's allele of Y (boy) is 100%, and the probability of the mother passing a c (colorblind) is 50%.

*Probability of a son being color blind* is 50% or ½.

If the question had asked, "what is the probability that they will have a color-blind child?" the analysis changes to determine the probability for all children (not just boys in the original question).

*Gametes* produced by the mother are C and c with a 50% probability each.

Affected child is a boy.

The probability that the father passes the X or Y gene to the offspring is 50%.

*Individual event* probabilities are multiplied to determine the overall probability: ½ × ½ = ¼

## 23. C is correct.

*Seminiferous tubules* are in the testes and are the site of sperm production.

*Spermatozoa* are the mature male gametes in many sexually reproducing organisms.

*Spermatogenesis* is how spermatozoa are produced from male primordial germ cells by mitosis and meiosis. Initial cells in this pathway are spermatogonia, which yield 1° spermatocytes by mitosis.

1° spermatocyte divides via meiosis into two 2° spermatocytes.

*Meiosis* converts 1° spermatocyte (2N) into four spermatids (1N).

2° spermatocytes complete meiosis by dividing into 2 spermatids as mature spermatozoa (i.e., sperm cells).

1° spermatocyte gives rise to two 2° spermatocytes that, by further meiosis, produce four spermatozoa.

*Seminiferous tubules* are in the testes and the location of meiosis and subsequent creation of gametes (e.g., spermatozoa for males or ova for females).

*continued...*

Epithelium of the tubule consists of *Sertoli cells,* whose primary function is to nourish the developing sperm cells through the stages of spermatogenesis.

*Sertoli* cells function as phagocytes, consuming the residual cytoplasm during spermatogenesis.

*Spermatogenic cells* differentiate between the Sertoli cells, which differentiate through meiosis into sperm cells.

## 24. E is correct.

*Primary oocytes* are arrested in meiotic prophase I from birth until ovulation within the ovaries.

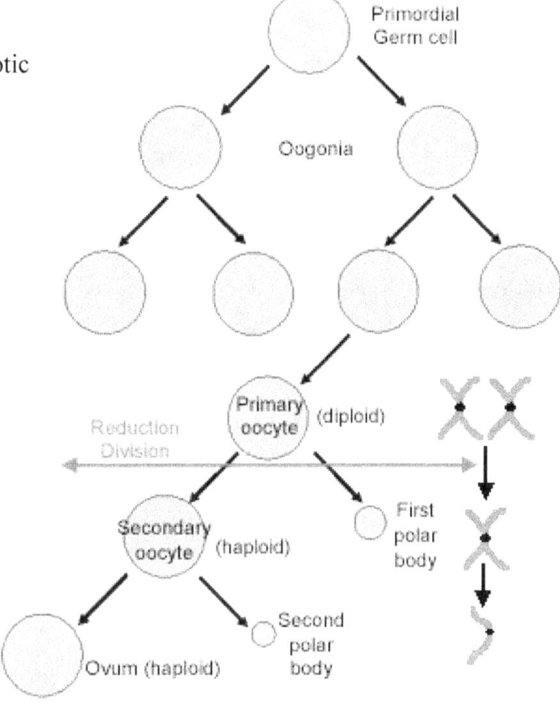


*Diploid primordial germ cell undergoes two rounds of meiosis, forming a haploid ovum and polar bodies*

## 25. C is correct.

A human cell after the first meiotic division is 1N with 2 chromatids.

## 26. D is correct.

*Genetic recombination* is when two DNA strand molecules exchange genetic information (i.e., a base composition within the nucleotides), resulting in new combinations of alleles (i.e., alternative forms of genes).

In eukaryotes, the natural process of genetic recombination during meiosis (i.e., the formation of gametes – eggs and sperm) results in genetic information passed to progeny.   *continued...*

Genetic recombination in eukaryotes involves pairing homologous chromosomes (i.e., a set of maternal and paternal chromosomes), which may involve nucleotide exchange between the chromosomes.

Information exchange may occur without physical exchange when a section of genetic material is duplicated without a change in the donating chromosome or by breaking and rejoining the DNA strands (i.e., forming new DNA molecules).

Mitosis may involve recombination, where two sister chromosomes form after DNA replication. New combinations of alleles are often not produced because the sister chromosomes are usually identical.

In meiosis and mitosis, recombination occurs between similar DNA molecules (homologous chromosomes or sister chromatids, respectively).

In meiosis, non-sister (i.e., same parent) homologous chromosomes pair with each other, and recombination often occurs between non-sister homologs. For somatic cells (mitosis) and gametes (meiosis), recombination between homologous chromosomes or sister chromatids is a common DNA repair mechanism.

## 27. C is correct.

*Primary oocytes* (2N cells) are arrested in prophase I of meiosis I until puberty. There is additional development into secondary oocytes that occurs prior to ovulation.

After ovulation, the oocyte is arrested in metaphase of *meiosis II until fertilization.*

## 28. B is correct.

*Pseudoautosomal regions* are named because any genes within them are inherited, like autosomal genes.

Pseudoautosomal regions allow males to pair and segregate X and Y chromosomes during meiosis.

Males have two copies of genes: one in the pseudoautosomal region of their Y chromosome and the other in their X chromosome's corresponding portion.

Typical females possess two copies of pseudoautosomal genes; each X chromosome contains pseudoautosomal regions.

Crossing over (during prophase I) between the X and Y chromosomes is usually restricted to the pseudoautosomal regions.

Pseudoautosomal genes exhibit an *autosomal*, rather than sex-linked, inheritance pattern.

Females can inherit an allele initially on the Y chromosome of their father, and males can inherit an allele initially on the X chromosome of their father.

**29. E is correct.**

*Primary spermatocyte* completes the synthesis (S) phase of interphase, not the first meiotic division, and remains diploid (2N) with 46 chromosomes (i.e., 23 pairs).

**30. B is correct.**

*Chromosomes* (not chromatids) *segregate* during mitosis to produce identical somatic (body) 2N (diploid) cells from the parental cell.

Meiosis is the process for germline cells (i.e., egg, sperm).

Mutations are inheritable changes in the cell's genetic (DNA) material; recombination occurs during prophase I (at the chiasma of the tetrad) of meiosis.

Homologous chromosomes segregate during meiosis I

During meiosis I, homologous chromosomes separate.

Subsequently, during meiosis II, the sister chromatids separate to produce four 1N (haploid) products, each with half the number of chromosomes as the original cell.

**31. E is correct.**

In females, secondary oocytes are haploid (i.e., single copies of 23 chromosomes, each with pair of chromatids).

*Secondary oocytes* do not complete meiosis II (i.e., haploid with a single chromatid) until fertilized by sperm.

Each month during puberty, one primary oocyte (i.e., diploid, each with a pair of chromatids) completes meiosis I to produce a secondary oocyte (1N) and a polar body (1N).

1N secondary oocyte (i.e., 23 chromosomes, each with a pair of chromatids) is expelled as an ovum from the follicle during ovulation, but meiosis II does not occur until fertilization.

*Menarche* is marked by the onset of menstruation and signals the possibility of fertility.

*Menstruation* is the sloughing of the *endometrial lining* of the uterus during the monthly hormonal cycle.

*Menopause* is when menstruation ceases.

*continued...*

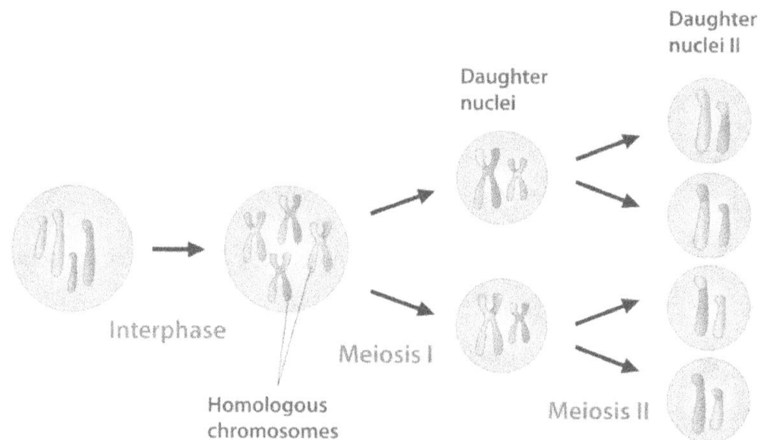



*Chromosomes replicate in S phase of mitosis to produce homologous chromosomes. Meiosis I daughter cells are diploid (2N), while the gametes formed from meiosis II (reductive phase) are haploid (1N).*

## 32.  D is correct.

*Klinefelter syndrome* describes the symptoms from additional X genetic material in males.

*Turner syndrome* is a condition in females with a single X (monosomy X) chromosome.

*XYY syndrome* is a genetic condition in which a human male has an extra male (Y) chromosome, giving 47 chromosomes instead of 46. XYY is not inherited but usually occurs during the formation of sperm cells.

*Nondisjunction* errors during anaphase II (meiosis II) result in sperm with an extra Y chromosome.

If an atypical sperm contributes to genetic makeup, the child has an extra Y chromosome in each somatic cell.

*Triple X syndrome* is not inherited but usually occurs during the formation of gametes (e.g., ovum and sperm) because nondisjunction in cell division results in reproductive cells with additional chromosomes.

During cell division, errors (non-disjunction) can result in gametes with additional chromosomes.

An egg or sperm may gain an extra X chromosome due to non-disjunction, and if one gamete contributes to the zygote, the child will have an extra X chromosome in each cell.

## 33. C is correct.

In males, diploid spermatogonia (2N) cells undergo meiosis I to produce 1° diploid spermatocyte (2N), which undergoes meiosis II to yield 2° haploid spermatocytes (1N).

2° spermatocytes undergo meiosis II to produce four spermatids (1N) that mature into spermatozoa (i.e., sperm).

In females, 1° oocyte is (2N), whereby 2° oocyte undergoes the second meiotic division to produce two (1N) cells – a mature oocyte (i.e., ovum – female gamete) and a polar body.

*Primary oocyte* is diploid (2N).

During fertilization, a (1N) ovum and a (1N) sperm fuse to produce a (2N) zygote.

*Spermatogonium* is a diploid (2N) cell.

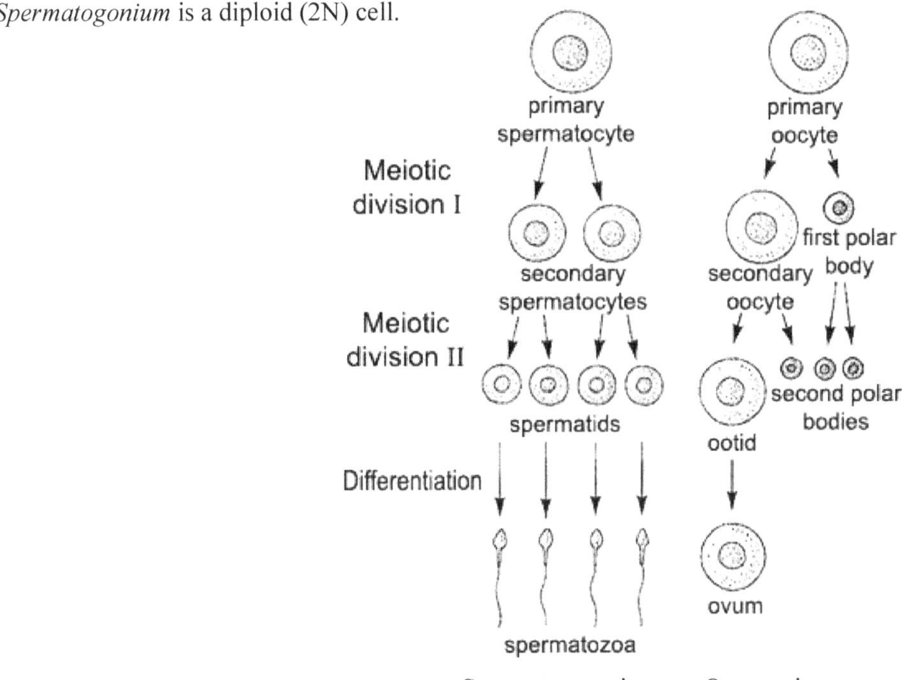


*Diploid primary oocytes and oocytes undergo two rounds of meiosis forming haploid spermatozoa and ovum*

## 34. D is correct.

*Probability* that a child will be male (or female) is ½.

Each event is *independent*; the first (or second or third) child does *not* affect future events.

Probability for 4 children is:

$$\frac{1}{2} \times \frac{1}{2} \times \frac{1}{2} \times \frac{1}{2} = 1/16$$

Same probability if first and third are female, with second and fourth male (or combination).

## 35. D is correct.

*Unequal division of cytoplasm* occurs during the meiotic process of oogenesis (i.e., production of an egg cell).

Meiotic divisions occur in two stages.

*First stage* produces precursor (2N) cells when a daughter cell (precursor egg) receives most of the cytoplasm; the other is devoid of sufficient cytoplasm (but genetically identical) and a nonfunctioning polar body.

*Second division of oogenesis* is when the large (2N) daughter cell divides again, and again one (1N) daughter cell receives the most cytoplasm (i.e., putative *ovum*).

The other daughter eggs become small nonfunctional (1N) *polar bodies.*

Polar bodies (from meiosis I) may divide (via meiosis II) to form two nonfunctional (1N) polar bodies.

The result is a potential of four (1N) cells.

However, only one—the one with a greater amount of cytoplasm during each meiotic division—becomes a functional egg (i.e., ovum) cell along with (up to) three (1N) polar bodies.

*Bacterial cells* divide for reproduction, and cytoplasmic division is equal.

*Mitosis of kidney* cells distributes cytoplasm equally.

*Spermatogenesis* is when one (2N) precursor cell forms four functional (1N) sperm cells because both meiosis divisions are equal and contain an equal amount of cytoplasm.

## 36. B is correct.

*Dyneins* are motor proteins (or *molecular motors*) using ATP to perform movement as *retrograde transport.*

Dynein transports cellular contents by *walking* along cytoskeletal microtubules towards the *minus-end* of the microtubules, usually oriented towards the cell center (i.e., *minus-end directed motors.*)

*Kinesins* are motor proteins that move toward the *plus-end* of the microtubules (i.e., *plus-end directed motors*).

*Ovulation* is the phase of the female menstrual cycle. A partially mature ovum that has yet to complete meiosis II is released from the ovarian follicles into the Fallopian tube (i.e., oviduct).

*continued...*

After ovulation, the egg can be fertilized by sperm during the *luteal phase.*

*Ovulation* is determined by circulating hormone levels and is not affected by a defect in dynein motor proteins.

*Lungs* require cilia to remove bacteria and other particulates.

*Kartagener's syndrome males* would be *infertile due to sperm immobility.*

*Ova* would not typically enter the Fallopian tubes (i.e., oviduct) because of the lack of cilia, which would cause an increased risk of ectopic pregnancy.

E: *epithelium lining* the tube of the middle ear is ciliated.

Beat is directed from the cavity to the pharynx, which clears mucus and pathogens.

Therefore, interference with the ciliary function would increase middle ear infections.

## 37.  E is correct.

*DNA replication* occurs during the synthesis (S) phase of interphase (i.e., G1, S, G2) to form *sister chromatids* joined by a centromere.

*Replication* occurs during the cell cycle *S* phase (i.e., interphase).

*Transcription* (in the nucleus) and *translation* (in the cytoplasm) occur during the S phase.

## 38.  C is correct.

*Progesterone* is a steroid hormone involved in the female menstrual cycle, pregnancy (supports *gestation*), and embryogenesis.

*Progesterone levels* are relatively low in women during the pre-ovulatory phase of the menstrual cycle, rise after ovulation, and are elevated during the luteal phase.

During pregnancy, human chorionic gonadotropin (HCG) is released, maintaining the corpus luteum and allowing it to maintain progesterone levels.

At 12 weeks, the placenta begins to produce progesterone instead of the corpus luteum – this process is the *luteal-placental shift.*

After the luteal-placental shift, progesterone levels start to rise further.

Progesterone levels are very low after delivery of the placenta and during lactation.

Progesterone levels are low in children and postmenopausal women.

Adult males have levels like those in women during the follicular phase of the menstrual cycle.

**39.  A is correct.**

During meiosis, gamete (e.g., ovum and sperm) reduces its genetic component from diploid (2N) to haploid (1N) with half the typical chromosome number for somatic (i.e., body cells).

When a haploid egg and sperm unite, they form a diploid zygote.

Ova contains an X chromosome, while sperm contains an X or a Y chromosome.

*Gametes form* in meiosis (i.e., two reduction divisions) without intervening chromosome replication.

During prophase I of meiosis I, *tetrads* form, and *sister chromatids* (i.e., a chromosome replicated in a prior S phase) undergo homologous recombination as *crossing over*.

*Crossing over* increases genetic variance within the progeny and is a driving force in the evolution of a species.

**40.  E is correct.**

*Meiosis* is cell division in sexually reproducing eukaryotes (animals, plants, and fungi), whereby the chromosome number is reduced by half, resulting in four genetically distinct haploid daughters.

*DNA replication* is followed by mitosis and *two rounds of cell division* in meiosis to produce four (1N) cells.

Two rounds of meiotic divisions are Meiosis I and Meiosis II.

Meiosis I is a *reductive division*, as the cells are reduced from diploid (2N) to haploid (1N).

Meiosis II (only G phase separates I and II) is an *equational division*, as the cells begin and end as haploids.

**41.  D is correct.**

*Crossing over* occurs during prophase I of meiosis.

During prophase I, the *chromatin condenses* into chromosomes, the centrioles migrate to the poles, the spindle fibers begin to form, and the nucleoli and nuclear membrane disappear.

*Homologous chromosomes* physically pair and intertwine in the process of *synapsis*.

Prophase I: *chromatids* (i.e., a strand of the replicated chromosome) of *homologous chromosomes* break and exchange equivalent pieces of DNA via *crossing over*.

Metaphase I: *homologous pairs align* at the equatorial plane, and each pair attaches to a separate spindle fiber by its *kinetochore* (i.e., protein collar around the centromere of the chromosome).

*continued...*

Anaphase I: *homologous chromosomes separate* and are pulled by the spindle fibers to opposite cell poles.

*Disjunction* (i.e., separation of homologous chromosomes) is essential for segregation, as described by Mendel.

Telophase I: *nuclear membrane forms* around each new nucleus, and chromosomes consist of sister chromatids joined at the centromere.

Cell divides into two daughter cells, each receiving a haploid nucleus (1N) of chromosomes.

*Interkinesis* is a brief period between the two reduction cell divisions of meiosis I and II, during which the chromosomes partially uncoil.

## 42.  A is correct.

In prophase I, *tetrads form*, genetic recombination occurs, and the spindle apparatus forms.

Chromosomes migrate to the poles of the cell during anaphase from the *splitting of the centromere*.

## 43.  E is correct.

*Spermatogenesis* and *oogenesis* are *gametogenesis*.

Haploid (1N) gametes (i.e., ova and sperm) are produced by diploid (2N) cell reductive divisions (i.e., meiosis).

*Spermatogenesis* occurs in the gonads, whereby the cytoplasm equally divides during meiosis by producing four viable sperm 1N cells.

*Oogenesis* occurs in the gonads, whereby the cytoplasm divides unequally.

One 1N ovum (e.g., egg) receives the bulk of the cytoplasm and (up to) three additional 1N polar bodies.

*Polar bodies* contain a 1N genome (like sperm and egg) but lack sufficient cytoplasm for a viable gamete.

*continued...*

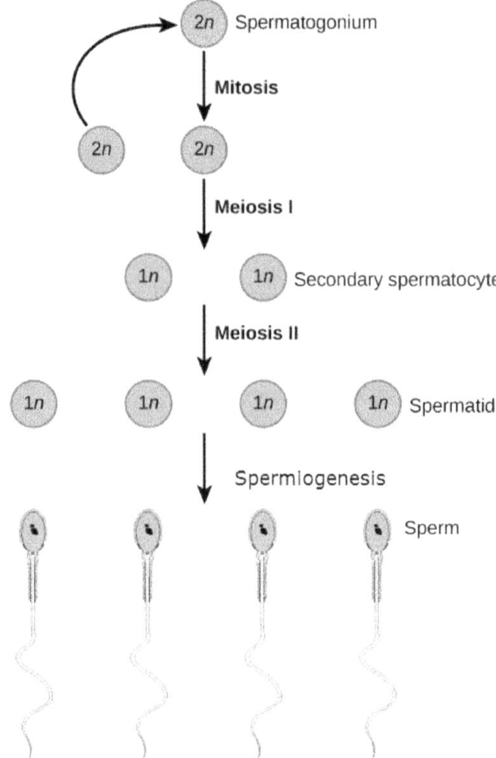



*Diploid primary oocytes and oocytes undergo two rounds of meiosis forming haploid spermatozoa and ovum*

### 44. B is correct.

*Barr body* is an *inactive X chromosome* in a female somatic cell rendered inactive by *lyonization* for species in which sex is determined by the Y chromosomes (i.e., humans and other species).

*Lyon hypothesis* states that cells with multiple X chromosomes inactivate all but one X randomly during mammalian embryogenesis.

In humans with more than one X chromosome, the number of Barr bodies visible at interphase is one less than the total number of X chromosomes.

For example, a man with Klinefelter syndrome of 47, XXY karyotype, has a single Barr body.

A woman with a 47, XXX karyotype has two Barr bodies.

*Polar bodies* form in females as 1N, nonfunctional cells during meiosis.

*continued...*

*Meiosis* is a two-stage process whereby a 2N cell undergoes a reduction division to form two 1N cells

Each 1N cell forms two 1N cells for four 1N cells from a parental 2N germ cell in the *second meiotic* division.

For sperm cells, four functional, unique 1N gamete sperm cells form.

For egg cells, the first meiotic division involves an *unequal division of cytoplasm* that results in one large cell and one small cell (i.e., the first polar body).

The large cell divides (second meiotic division) to generate *one functional egg* (1N ovum containing most of the cytoplasm) and another polar body.

*First polar body* divides again to form two other polar bodies. One large ovum (i.e., functional egg cell) and three polar bodies form during a female meiotic division.

Polar bodies form in mitosis, which is how somatic (i.e., body cells) cell divisions occur; equal cells do not form gametes (i.e., germline cells).

## 46.  E is correct.

*Turner syndrome* (i.e., 45, X) describes several conditions in females, of which monosomy X (i.e., the absence of the entire sex (X) chromosome, the Barr body) is the most common.

Turner syndrome is a chromosomal abnormality when all or part of a sex chromosome is absent or abnormal.

## 47.  D is correct.

Dolly, the sheep, was cloned in 1996 by fusing the *nucleus* from a mammary gland cell of a Finn Dorset ewe into an *enucleated egg cell* from a Scottish Blackface ewe.

During gestation, Dolly was carried to term in the uterus of another Scottish Blackface ewe.

She was a genetic copy of the Finn Dorset ewe from the somatic mammary gland cell.

Dolly's creation showed that somatic (i.e., body cell) cell DNA in a differentiated (*vs. pluripotent* or *totipotent*) could be induced through nuclear transfer (i.e., transplantation) to expand developmental fate (e.g., like a germ cell) in progression from a zygote.

*Notes for active learning*

## Practice Questions: Development

1. Which statement is true regarding respiratory exchange during fetal life?

    A. Respiratory exchanges are made through the placenta
    B. Since lungs develop later in gestation, a fetus does not need a mechanism for respiratory exchange
    C. The respiratory exchange is made through the ductus arteriosus
    D. The respiratory exchange is not necessary
    E. The respiratory exchange is made through the ductus venosus

2. When each cell in an early stage of embryonic development still has the ability to develop into a complete organism, it is known as:

    A. gastrulation
    B. blastulation
    C. determinate cleavage
    D. indeterminate cleavage
    E. none of the above

3. Mutations in *Drosophila*, resulting in the transformation of one body segment into another, effect:

    A. maternal-effect genes
    B. homeotic genes
    C. execution genes
    D. segmentation genes
    E. gastrulation genes

4. Which structure is developed from embryonic ectoderm?

    A. connective tissue
    B. bone
    C. rib cartilage
    D. epithelium of the digestive system
    E. hair

5. The placenta is a vital metabolic organ made from a contribution by the mother and fetus. The portion of the placenta contributed by the fetus is:

    A. amnion
    B. yolk sac
    C. umbilicus
    D. chorion
    E. none of the above

**6.** What occurs when an embryo lacks human chorionic gonadotropin (hCG) synthesis?

**A.** The embryo does not support the maintenance of the corpus luteum

**B.** The embryo increases the production of progesterone

**C.** The embryo develops immunotolerance

**D.** The placenta forms prematurely

**E.** Conception does not occur

**7.** Which germ layer gives rise to smooth muscle?

**A.** ectoderm

**B.** epidermis

**C.** mesoderm

**D.** endoderm

**E.** hypodermis

**8.** In early embryogenesis, the most critical morphological change in the development of cellular layers is:

**A.** epigenesis

**B.** gastrulation

**C.** conversion of morula to blastula

**D.** acrosomal reaction

**E.** fertilization membrane

**9.** During the first eight weeks of development, all events occur EXCEPT:

**A.** myelination of the spinal cord

**B.** presence of all body systems

**C.** formation of a functional cardiovascular system

**D.** beginning of ossification

**E.** limb buds appear

**10.** Which structures derive from the ectoderm germ layers?

**A.** blood vessels, tooth enamel, and epidermis

**B.** heart, kidneys, and blood vessels

**C.** nails, epidermis, and blood vessels

**D.** epidermis and adrenal cortex

**E.** epidermis and neurons

**11.** A diagram of the blastoderm that identifies regions from which specific adult structures are derived is a:

**A.** phylogenetic tree

**B.** pedigree diagram

**C.** fate map

**D.** linkage map

**E.** lineage diagram

**12.** Which tissue is the precursor of long bones in the embryo?

**A.** hyaline cartilage

**B.** fibrocartilage

**C.** dense fibrous connective tissue

**D.** elastic cartilage

**E.** costal cartilage

**13.** What changes are observed in cells as development proceeds?

**A.** Cytoskeletal elements involved in forming the mitotic spindle

**B.** Energy requirements of each cell

**C.** Genetic information that was duplicated with each round of cell division

**D.** Composition of polypeptides within the cytoplasm

**E.** Composition of nucleotides within the nucleus

**14.** During fertilization, what is the earliest event in the process?

**A.** sperm contacts the cortical granules around the egg

**B.** sperm nucleic acid enters the egg's cytoplasm to form a pronucleus

**C.** acrosome releases hydrolytic enzymes

**D.** sperm contacts the vitelline membrane around the egg

**E.** hyaluronic acid forms a barrier to prevent polyspermy

**15.** The embryonic ectoderm layer gives rise to all the following EXCEPT:

**A.** eyes

**B.** fingernails

**C.** integument

**D.** nervous system

**E.** blood vessels

**16.** All are correct matches of a fetal structure with what it becomes at birth, EXCEPT:

**A.** ductus venosus—ligamentum venosum

**B.** umbilical arteries—medial umbilical ligament

**C.** foramen ovale—fossa ovalis

**D.** ductus arteriosus—ligamentum teres

**E.** all choices are correct

**17.** All statements regarding gastrulation are true, EXCEPT:

**A.** the primitive gut that results from gastrulation is the archenteron

**B.** for amphibians, gastrulation is initiated at the gray crescent

**C.** after gastrulation, the embryo consists of two germ layers of endoderm and ectoderm

**D.** for amphibians, blastula cells migrate during gastrulation by an invagination region of the blastopore

**E.** the mesoderm develops as the third primary germ layer during gastrulation


**18.** A homeobox is a:

**A.** protein involved in the control of meiosis

**B.** transcriptional activator of other genes

**C.** DNA binding site

**D.** sequence responsible for post-transcriptional modifications

**E.** sequence that encodes for a DNA binding motif


**19.** After the gastrulation's completion, the embryo undergoes the following:

**A.** cleavage

**B.** blastulation

**C.** blastocoel formation

**D.** neurulation

**E.** dedifferentiation


**20.** Shortly after implantation:

**A.** trophoblast forms two distinct layers

**B.** embryo undergoes gastrulation within 3 days

**C.** maternal blood sinuses bathe the inner cell mass

**D.** myometrial cells cover and seal off the blastocyst

**E.** the umbilical cord forms within 48 hours


**21.** What is the anatomical connection between the placenta and embryo?

**A.** chorion

**B.** umbilical cord

**C.** endometrium

**D.** corpus luteum

**E.** vas deferens

**22.** Which statement correctly illustrates the principle of induction during vertebrate development?

- **A.** Neural tube develops into the brain, the spinal cord, and the nervous system
- **B.** Secretion of TSH stimulates the release of thyroxine hormone
- **C.** Ectoderm develops into the nervous system
- **D.** Neurons synapse with other neurons via neurotransmitters
- **E.** Presence of a notochord beneath the ectoderm results in the formation of a neural tube

**23.** The acrosomal reaction by the sperm is:

- **A.** a transient voltage change across the vitelline envelope
- **B.** the jelly coat blocking penetration by multiple sperms
- **C.** the consumption of yolk protein
- **D.** hydrolytic enzymes degrading the plasma membrane
- **E.** the inactivation of the sperm acrosome

**24.** Which primary germ layer gives rise to the cardiovascular system, bones, and skeletal muscles?

- **A.** endoderm
- **B.** blastula
- **C.** mesoderm
- **D.** ectoderm
- **E.** gastrula

**25.** The dorsal surface cells of the inner cell mass form:

- **A.** notochord
- **B.** placenta
- **C.** one of the fetal membranes
- **D.** structure called the embryonic disc
- **E.** primitive streak

**26.** All occur in a newborn immediately after birth, EXCEPT:

- **A.** the infant completely stops producing fetal hemoglobin
- **B.** resistance in the pulmonary arteries decreases
- **C.** pressure in the left atrium increases
- **D.** pressure in both the inferior vena cava and the right atrium increase
- **E.** ductus arteriosus constricts

**27.** Homeobox sequences are present in:

**A.** introns

**B.** exons

**C.** 3'-untranslated regions

**D.** 5'-untranslated regions

**E.** exon-intron boundaries

**28.** After 30 hours of incubation, the ectoderm tissue within the gastrula differentiates into different specific tissues, which supports the conclusion that:

I. cells become either endoderm or mesoderm

II. cells contain a different genome from their parental cells

III. gene expression is altered

**A.** III only

**B.** I and III only

**C.** I only

**D.** I and II only

**E.** II and III only

**29.** What is the function of the yolk sac in humans?

**A.** forms into the placenta

**B.** secretes progesterone in the fetus

**C.** gives rise to blood cells and gamete-forming cells

**D.** stores embryonic waste

**E.** stores nutrients for the embryo

**30.** The trophoblast is mostly responsible for forming:

**A.** placental tissue

**B.** lining of the endometrium

**C.** allantois

**D.** archenteron

**E.** chorion

**31.** Certain lesions to the mesodermal embryonic primary germ layer may stimulate the development of *spina bifida*, a congenital fissure in the lower vertebrae. Besides the spinal column, what other structures would be affected by such lesions?

I. intestinal epithelium

II. skin and hair

III. blood vessels

IV. muscles

**A.** I and II only

**B.** II and III only

**C.** III and IV only

**D.** I, III and IV only

**E.** IV only

**32.** The homeotic genes encode for:

**A.** repressor proteins

**B.** transcriptional activator proteins

**C.** helicase proteins

**D.** single-strand binding proteins

**E.** restriction enzymes

**33.** Which cells give rise to muscles in a frog embryo?

**A.** neural tube

**B.** ectoderm

**C.** endoderm

**D.** mesoderm

**E.** notochord

**34.** Which is NOT involved in the implantation of the blastocyst?

**A.** Settling of the blastocyst onto the prepared uterine lining

**B.** Adherence of the trophoblast cells to the uterine lining

**C.** Phagocytosis by the trophoblast cells

**D.** Inner cell mass giving rise to primitive streak

**E.** Endometrium proteolytic enzymes produced by the trophoblast cells

**35.** During labor, which hormone stimulates contractions of uterine smooth muscle?

**A.** oxytocin

**B.** prolactin

**C.** luteinizing hormone

**D.** hCG

**E.** estrogen

**36.** In a chick embryo, specific ectoderm cells give rise to wing feathers, while others develop into thigh feathers or foot claws. The ectodermal cells that develop into wing feathers were transplanted to an area that develops into thigh feathers or feet claws. It was observed that the transplanted cells developed into claws. Which best explains the experimental results?

**A.** ectoderm cells possess positional information

**B.** destiny of the cells was already determined

**C.** ectoderm cells can develop into any tissue

**D.** the underlying mesoderm-induced cells

**E.** ectoderm released growth factors

**37.** The *slow block* to polyspermy is due to the following:

**A.** transient voltage changes across membranes

**B.** jelly coat blocking sperm penetration

**C.** consumption of yolk protein

**D.** formation of the fertilization envelope

**E.** inactivation of the sperm acrosome

**38.** Polyspermy in humans results in:

**A.** mitotic insufficiency

**B.** interruption of meiosis

**C.** nonviable zygote

**D.** multiple births

**E.** formation of multiple placentas

**39.** Mesoderm gives rise to:

**A.** intestinal mucosa

**B.** nerves

**C.** skin

**D.** lung epithelium

**E.** heart

**40.** Genomic imprinting is:

**A.** inactivation of a gene by interruption of its coding sequence

**B.** organization of molecules in the cytoplasm to provide positional information

**C.** DNA modification in gametogenesis that affects gene expression in the zygote

**D.** suppression of a mutant phenotype because of a mutation in a different gene

**E.** mechanism by which enhancers distant from the promoter can still regulate transcription

**41.** Comparing a developing frog embryo and an adult, cells of which have a greater translation rate?

**A.** Adult, because ribosomal production is more efficient in a mature organism

**B.** Adult, because a mature organism has more complex metabolic requirements

**C.** Embryo, because a developing organism requires more protein production than an adult

**D.** Embryo, because ribosomal production is not yet under regulatory control by DNA

**E.** Embryo, because the proteins must undergo more extensive post-translation processing

**42.** Which structure is the first to form during fertilization in humans?

**A.** blastula

**B.** morula

**C.** neural tube

**D.** ectoderm

**E.** zygote

**43.** It is impossible for sperm to be functional (i.e., able to fertilize the egg) until after:

I. they undergo capacitation

II. the tail disappears

III. they have been in the uterus for several days

IV. they become spermatids

**A.** I only

**B.** II and IV only

**C.** I and IV only

**D.** I and III only

**E.** IV only

**44.** Where is the deformity if a teratogen affects the endoderm development shortly after gastrulation?

**A.** nervous system

**B.** lens of the eye

**C.** liver

**D.** skeleton

**E.** connective tissue

**45.** Mutations that cause cells to undergo developmental fates of other cell types are:

**A.** heterochronic mutations

**B.** loss-of-function mutants

**C.** transection mutations

**D.** execution mutations

**E.** homeotic mutations

**46.** Each is true for the cells of an early gastrula's eye field, EXCEPT that they are:

**A.** terminally differentiated

**B.** competent

**C.** derived from the ectoderm layer

**D.** capable of becoming other ectoderm structures

**E.** undifferentiated ectoderm

**47.** What is the role of proteases and acrosin enzymes in reproduction?

   **A.** They degrade the nucleus of the egg and allow the sperm to enter

   **B.** They degrade the protective barriers around the egg and allow the sperm to penetrate

   **C.** They direct the sperm to the egg through chemotaxis messengers

   **D.** They neutralize the mucous secretions of the uterine mucosa

   **E.** They degrade the protective barriers around the sperm and allow the egg to penetrate


**48.** When does the human blastocyst implant in the uterine wall?

   **A.** about a week past fertilization       **C.** a few hours past fertilization

   **B.** at blastulation       **D.** at primary germ formation

                                     **E.** at primitive streak formation


**49.** Which primary layer develops into the retina of the eye?

      I. endoderm         II. ectoderm         III. mesoderm

   **A.** I only                  **C.** III only

   **B.** II only                **D.** II and III

                                     **E.** I and III


**50.** What changes must occur in a newborn's cardiovascular system after the infant takes its first breath?

   **A.** Ductus arteriosus constricts and is converted to the ligamentum arteriosum

   **B.** The urinary system is activated at birth

   **C.** Ductus venosus is severed from the umbilical cord, and visceral blood enters vena cava

   **D.** Foramen ovale between the atria of the fetal heart closes at the moment of birth

   **E.** Foramen ovale becomes the medial umbilical ligament


**51.** Early activation of the mammalian zygote nucleus may be necessary because:

   **A.** most developmental decisions are made under the influence of the paternal genome

   **B.** mammalian oocytes are too small to store molecules needed to support the cleavage divisions

   **C.** in gametogenesis, specific genes undergo imprinting

   **D.** mammals do not have maternal-effect genes

   **E.** none of the above

**52.** In reptiles, the aquatic environment necessary for the embryonic development of amphibians is replaced by:

- **A.** use of lungs instead of gills
- **B.** humid atmospheric conditions
- **C.** shells which prevent the escape of gas
- **D.** intrauterine development
- **E.** amniotic fluid

**53.** What would be expected to form from a portion of cells destined to become the heart if it was excised from an early gastrula and placed in a culture medium?

- **A.** undifferentiated mesoderm
- **B.** undifferentiated ectoderm
- **C.** differentiated endoderm
- **D.** undifferentiated endoderm
- **E.** differentiated ectoderm

**54.** During gestation, when can an ultrasound determine the sex of the fetus?

- **A.** at the midpoint of the first trimester
- **B.** about 18 weeks after fertilization
- **C.** at the end of the first trimester
- **D.** at the midpoint of the second trimester
- **E.** at the midpoint of the third trimester

**55.** How long is the egg viable and able to be fertilized after ovulation?

- **A.** 36-72 hours
- **B.** a full week
- **C.** up to 6 hours
- **D.** 24-36 hours
- **E.** 12-24 hours

**56.** Which stage of embryonic development has a hollow ball of cells surrounding a fluid-filled center?

- **A.** 3-layered gastrula
- **B.** blastula
- **C.** morula
- **D.** zygote
- **E.** 2-layered gastrula

**57.** Which statement describes the acrosome of a sperm cell?

   **A.** It contains nucleic acid

   **B.** It contains hydrolytic enzymes, which are released when the sperm encounters the jelly coat of the egg

   **C.** It fuses with the egg's cortical granules

   **D.** It functions to prevent polyspermy

   **E.** It is used for the motility of the sperm along the Fallopian tubes (oviducts)

**58.** Which statement about fertilization is correct?

   **A.** Most sperm cells are protected and remain viable once inside the uterus

   **B.** If estrogen is present, the pathway through the cervical opening is blocked from sperm entry

   **C.** The vagina's acidic environment destroys millions of sperm cells

   **D.** Spermatozoa remain viable for about 72 hours in the female reproductive tract

   **E.** The ovulated secondary oocyte is viable for about 72 hours in the female reproductive tract

**59.** All statements are correct regarding cleavage in human embryos, EXCEPT:

   **A.** blastomeres are genetically identical to the zygote

   **B.** holoblastic cleavage occurs in one portion of the egg

   **C.** morula is a solid mass of cells produced via cleavage of the zygote

   **D.** size of the embryo remains constant throughout the initial zygote cleavage

   **E.** none of the above

**60.** All statements are correct regarding cleavage in human embryos, EXCEPT:

   **A.** blastomeres are genetically identical to the zygote

   **B.** holoblastic cleavage occurs in one portion of the egg

   **C.** morula is a solid mass of cells produced via cleavage of the zygote

   **D.** size of the embryo remains constant throughout the cleavage of the zygote

   **E.** none of the above

*Notes or active learning*

*Notes or active learning*

## Detailed Explanations: Development

### Answer Key

| | | | | | |
|---|---|---|---|---|---|
| 1: A | 11: C | 21: B | 31: C | 41: C | 51: C |
| 2: D | 12: A | 22: E | 32: B | 42: E | 52: E |
| 3: B | 13: D | 23: D | 33: D | 43: A | 53: A |
| 4: E | 14: C | 24: C | 34: D | 44: C | 54: C |
| 5: D | 15: E | 25: E | 35: A | 45: E | 55: E |
| 6: A | 16: D | 26: D | 36: D | 46: A | 56: B |
| 7: C | 17: C | 27: B | 37: D | 47: B | 57: B |
| 8: B | 18: E | 28: A | 38: C | 48: A | 58: C |
| 9: A | 19: D | 29: C | 39: E | 49: B | 59: B |
| 10: E | 20: A | 30: A | 40: C | 50: A | 60: B |

### 1. A is correct.

*Respiratory exchanges* during fetal life occur through the *placenta.*

*Placenta* connects the developing fetus to the uterine wall and allows nutrient uptake, waste elimination, and gas exchange via the mother's blood supply.

*Umbilical cord* connects the developing embryo (or fetus) and the placenta.

During prenatal development, the umbilical cord is physiologically and genetically part of the fetus and contains two arteries (the umbilical arteries) and one vein (the umbilical vein).

*Umbilical veins* supply the fetus with oxygenated, nutrient-rich blood from the placenta.

Conversely, the fetal heart pumps deoxygenated, nutrient-depleted blood through the umbilical arteries to the placenta.

*Fetal circulatory systems* change at birth as the newborn uses its lungs.

After the infant's first breath, the newborn's cardiovascular system constricts the ductus arteriosus (i.e., connects the pulmonary artery to the aorta) and converts it to the *ligamentum arteriosum.* Resistance in the pulmonary blood vessels decreases, increasing blood flow to the lungs.

At birth, umbilical blood flow ceases, and blood pressure in the inferior vena cava decreases, causing a decrease in pressure in the *right atrium.*

*Left atrial pressure* increases due to increased blood flow from the lungs.

*continued...*

---

Increased left atrial pressure and decreased right atrial pressure cause *closure* of the *foramen ovale.*

*Ductus venosus,* which shunts blood from the left umbilical vein directly to the *inferior vena cava* to allow oxygenated blood from the placenta to bypass the liver, completely closes within three months after birth.

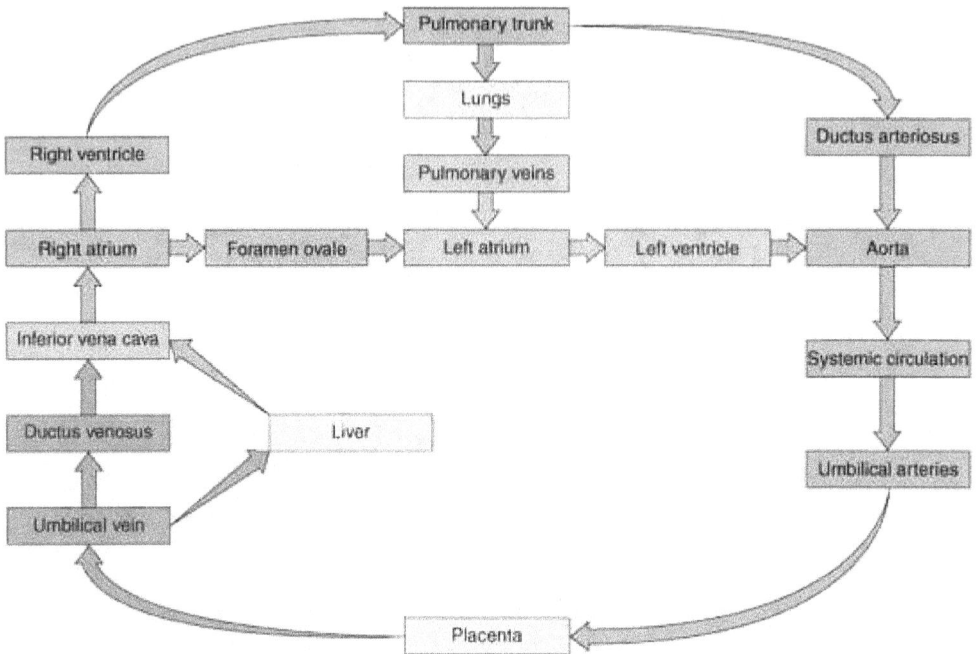


*Adult and fetal circulatory systems with ductus arteriosus and ductus venosus provide shunt pathways*

## 2. D is correct.

*Indeterminate cleavage* cells maintain the ability to develop into a complete organism.

*Zygote implantation* into the uterus causes cell migration transforming the blastula from a single-cell layer into a three-layered gastrula (i.e., ectoderm, mesoderm, and endoderm).

*Blastulation* begins when the morula develops the blastocoel as a fluid-filled cavity that (by the fourth day) becomes the blastula as a hollow sphere of cells.

*Determinate cleavage* results in cells whose differentiation potential is determined early in development.

## 3. B is correct.

Mutations in *Drosophila*, transforming one body segment into a different one, are in homeotic genes.

*Homeotic genes* encode the related homeodomain proteins involved in developmental patterns and sequences.

Homeotic genes influence the development of specific structures in plants and animals, such as the Hox and ParaHox genes, which are essential for segmentation.

Homeotic genes determine where, when, and how body segments develop.

For example, alterations in homeotic genes in laboratory flies cause changes in patterns of body parts, sometimes producing dramatic effects, such as legs growing in place of antennae or an extra set of wings.

Hox genes are *homeotic transcription factors* crucial for controlling the body plan along the *anterior-posterior axis* (i.e., craniocaudal axis) and specify the segment identity of tissues within the embryo.

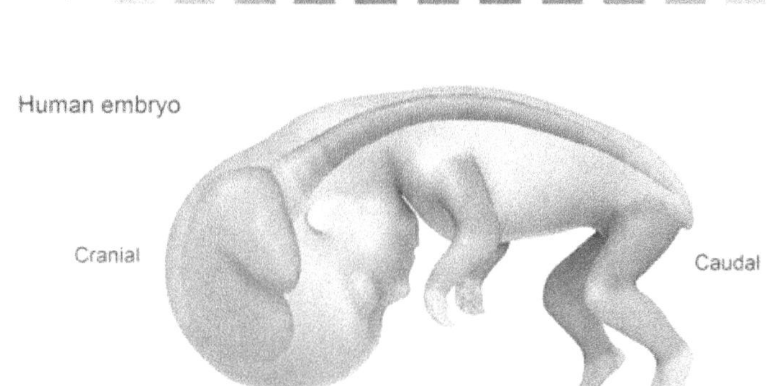


*HOX genes encode homeodomain proteins as developmental regulators of the anterior-posterior axis*

*Homeodomain* is a protein structural domain that binds DNA or RNA and is a transcription factor (i.e., facilitates polymerase binding and activation).

*Homeodomain proteins* fold with a 60-amino acid helix-turn-helix structure in which the short loop regions connect three alpha-helices.

*N-terminus* of the two helices is *antiparallel,* and the longer C-terminal helix is roughly perpendicular to the axes established by the first two.

Third helix functions as a *transcription factor* and interacts directly with DNA.

 

**4.  E is correct.**

*Ectoderm* gives rise to hair, nails, skin, brain, and nervous system.

Ectoderm forms "*outer linings*," including the epidermis (i.e., outermost skin layer) and hair.

Ectoderm is the precursor to *mammary glands*, *central* (CNS), and *peripheral nervous systems* (PNS).

*Connective tissue* is derived from *mesoderm*.

*Rib cartilage* is derived from *mesoderm*.

*Epithelium* of the digestive system is derived from *endoderm*.

**5.  D is correct.**

*Chorion* is a membrane between the developing fetus and mother in humans and most mammals.

Chorion consists of two layers:

> *outer layer* formed by the *trophoblast*

> *inner layer* formed by the *somatic mesoderm* (in contact with the amnion).

**6.  A is correct.**

The embryo does *not* support the maintenance of the *corpus luteum* when an embryo lacks the synthesis of human chorionic gonadotropin (hCG).

*Human chorionic gonadotropin* interacts with a receptor on the ovary, producing progesterone secretion.

*Progesterone* creates a thick lining of blood vessels and capillaries in the uterus to sustain the fetus.

Thus, the corpus luteum is supported during pregnancy.

**7.  C is correct.**

*Mesoderm* develops into the circulatory, musculoskeletal, and excretory systems, outer coverings of internal organs, gonads, and various types of muscle tissue.

*Ectoderm* develops into the brain and nervous system, hair and nails, lens of the eye, inner ear, sweat glands, the lining of the nose and mouth, and skin epidermis.

*Endoderm* develops into the epithelial lining of the digestive tract, respiratory tracts, lining of the liver, bladder, pancreas, thyroid, and alveoli of the lungs.

*Epidermis* is not an embryonic germ layer but the layer of the skin covering the dermis.

## 8. B is correct.

*Gastrulation* is a phase early in animal embryonic development, during which the single-layered blastula is reorganized into a trilaminar (*three-layered*) structure of the *gastrula.*

Three germ layers are:

*ectoderm, mesoderm,* and *endoderm.*

Gastrulation occurs after cleavage, formation of the blastula, and primitive streak, followed by *organogenesis.*

Organogenesis is when individual organs develop within the newly formed germ layers.

Following gastrulation, cells are organized into sheets of connected cells (e.g., epithelial) or mesh of isolated cells (i.e., *mesenchyme*).

Each germ layer gives rise to specific tissues and organs in the developing embryo.

*Ectoderm* gives rise to the epidermis and other tissues that will form the nervous system.

*Mesoderm* is between the ectoderm and the endoderm.

Mesoderm gives rise to somites that form muscle, cartilage of the ribs and vertebrae, dermis, notochord, blood and blood vessels, bone, and connective tissue.

*Endoderm* gives rise to epithelium of the respiratory system, digestive system, and organs associated with the digestive system (e.g., liver and pancreas).

*Gastrulation* occurs when a blastula of one layer folds inward and enlarges to create the *three primary germ layers* (i.e., endoderm, mesoderm, and ectoderm).

*Archenteron* gives rise to the digestive tube.

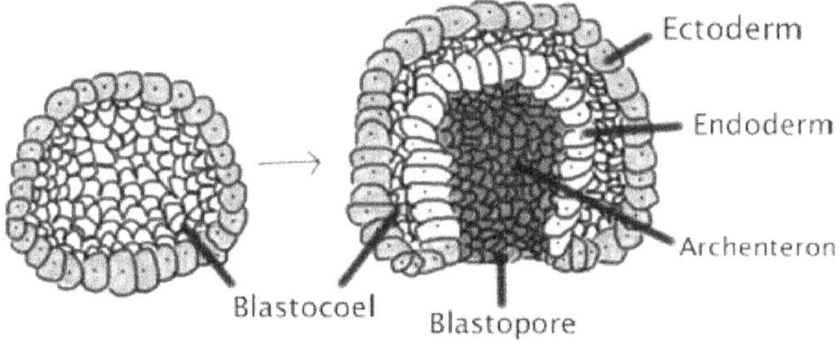


*Blastula (left) becomes the gastrula (right) with the formation of three primary germ layers*

**9.  A is correct.**

During the first *eight weeks* of development, myelination of the spinal cord does not occur.

*Myelination* of the *corticospinal tract* starts around 36 weeks and finishes around two years.

Most developmental milestones before two years are due to *myelination of nerve fibers*.

**10.  E is correct.**

*Ectoderm* germ layers give rise to the skin epidermis and the nervous system.

*Endoderm* germ layer gives rise to the lining of the digestive system, its associated glands, organs (e.g., liver and pancreas), and the lungs.

*Mesoderm* gives rise to most organs and systems, including excretory, reproductive, muscular, skeletal, and circulatory systems. If tissue is not specifically endoderm or ectoderm, it is likely to *be mesoderm derived*.

The incorrect choices are *mesoderm-derived* tissues or structures.

**11.  C is correct.**

*Fate mapping* traces the embryonic origin of tissues in adults by establishing the correspondence between individual cells (or groups of cells) at one stage of development and their progeny at later stages.

When conducted at single-cell resolution, this process is *cell lineage tracing*.

*Phylogenetic tree* (i.e., an evolutionary tree) is a branching diagram showing the inferred evolutionary relationships (phylogeny) among species on similarities and differences in physical or genetic characteristics.

Taxa joined in the tree are implied to have descended from a *common ancestor*.

*Pedigree diagram* shows the occurrence and appearance (i.e., *phenotypes*) of a gene in ancestors.

A pedigree presents family information in an easily readable chart.

Pedigrees use *standardized symbols*: *squares* represent males and *circles* females.

Pedigree construction is *family history*; details about an earlier generation may be uncertain as memories fade.

For unknown sex, a diamond is used.

Phenotype in question is represented by a filled-in (darker) symbol.

A shaded dot inside a symbol or a half-filled symbol indicates heterozygotes.

*continued…*

*Linkage map* is a genetic diagram of an experimental population that shows the relative position of its known genes or genetic markers based on recombination frequency rather than the physical distance on chromosomes.

*Genetic linkage* is the tendency of genes proximal on a chromosome to be inherited together during meiosis.

Genes whose loci (i.e., position) are closer are *less likely* to be separated onto different chromatids during chromosomal crossover (i.e., during prophase I) and are genetically *linked.*

## 12.  A is correct.

*Hyaline cartilage* tissue is the precursor of long bones in the embryo.

*Fibrous cartilage* is in intervertebral discs.

*Dense fibrous connective tissue* generally forms *tendons* and *ligaments.*

*Elastic cartilage* provides internal support and recoil to the external ear and epiglottis.

## 13.  D is correct.

*Gene expression* changes as development proceeds; expressed genes *encode proteins.*

During development, cells change their ability to respond to signals from tissues and induce changes in other cells.

These changes during development are inherited by daughter cells.

*Microtubules* are involved in mitosis but do not change during development.

*Somatic cells* have the same genes (exceptions include gametes and B and T cells).

## 14.  C is correct.

*Acrosome*s are at the tip of the sperm and contain specialized secretory molecules.

*Acrosomal reaction* is a signaling cascade involving the glycoproteins on the egg's surface.

*Acrosin* digests the zona pellucida and membrane of the oocyte.

Part of the sperm cell membrane fuses with the egg cell's membrane, and the contents of the head enter the egg fused with the plasma membrane.

This allows the sperm to release its degradation enzymes, penetrate the egg's tough coating, and bind and fuse.

*Pronucleus* is the nucleus of a sperm or an egg after a sperm enters the ovum but before pronucleus fusion.

**15. E is correct.**

*Three primary germ layers* form during gastrulation in embryogenesis: *ectoderm, mesoderm, and endoderm.*

*Ectoderm* is the external germ layer giving rise to the skin, fingernails, and nervous system (including the eye).

*Mesoderm* is the middle germ layer and gives rise to most organ systems, including the musculoskeletal, cardiovascular, reproductive, and excretory systems.

*Endoderm* is the innermost germ layer and gives rise to the gallbladder, liver, pancreas, and epithelial lining of luminal structures and accessory digestive organs.

**16. D is correct.**

*Ductus arteriosus* is a blood vessel connecting the pulmonary artery to the proximal descending aorta in the developing fetus.

Ductus arteriosus allows most blood from the right ventricle to bypass the fetus's fluid-filled non-functioning lungs.

*Ductus arteriosus* becomes the *ligamentum arteriosum* upon closure at birth.

Ligamentum teres (i.e., round ligament) refers to several structures:

> ligamentum teres *uteri* (i.e., round ligament of the uterus)

> ligamentum teres *hepatis* (round ligament of the liver)

> ligamentum teres *femoris* (i.e., ligament of the head of the femur)

**17. C is correct.**

*Blastula invagination* forms *mesoderm* as the third primary germ layer during gastrulation.

**18. E is correct.**

*Homeotic genes* encode homeodomain proteins involved in developmental patterns.

*Homeodomain* is a protein structural domain that binds DNA or RNA and is often a transcription factor.

*Homeodomain protein folding* consists of a 60-amino acid helix-turn-helix structure in which short loop regions connect three alpha-helices.

N-terminus of the two helices is antiparallel, and the longer C-terminal helix is perpendicular to the axes established by the first two.

*Third helix* functions as a *transcription factor and interacts directly with DNA.*

**19.  D is correct.**

*Neurulation* and *organogenesis* follow gastrulation.

*Mitosis* continues throughout development, but *cleavage* is a term reserved for *the first few cell divisions* when the *zygote* becomes the *morula*.

*Cleavage* has no growth occurring, and the morula is the same approximate size as the zygote.

*Blastula formation* precedes gastrulation.

*Blastocoel* is the fluid-filled central region of a blastula and forms early after fertilization when the zygote divides into many cells.

E: once *differentiated* (i.e., morphological, and biochemical distinct), a cell does not reverse this differentiation process unless it is a cancerous cell.

**20.  A is correct.**

*Trophoblast cells* give rise to the outer layer of a blastocyst, provide nutrients to the embryo, and develop into a large part of the placenta.

Trophoblasts form during the first stage of pregnancy and the first cells differentiate from the fertilized egg.

**21.  B is correct.**

*Placenta* connects the developing fetus to the uterine wall and allows nutrient uptake, waste elimination, and gas exchange via the mother's blood supply.

*Umbilical cord* connects the developing embryo (or fetus) and the placenta.

During prenatal development, the *umbilical cord* is physiologically and genetically part of the fetus and contains two arteries (the *umbilical arteries*) and one vein (the *umbilical vein*).

*Umbilical veins* supply the fetus with oxygenated, nutrient-rich blood from the placenta.

*Fetal heart* pumps deoxygenated, nutrient-depleted blood in umbilical *arteries* to the placenta.

**22.  E is correct.**

For vertebrates, *induction* is how a group of cells cause *differentiation* in another group of cells.

For example, cells that form the notochord induce the formation of the neural tube.

For example, induction in vertebrate development is eye formation, where the optic vesicles induce the ectoderm to thicken and form the lens placode (i.e., thickened portion of ectoderm becomes the lens), which induces the optic vesicle to form the optic cup, which induces the lens placode to form the cornea.                                    *continued...*

                                    

*Neural tube* does develop into the nervous system but is not induced by another group of cells or tissue.

*Thyroxin stimulating hormone* (TSH) stimulates thyroxine secretion but is not induction.

*Neurons synapse* with other neurons via neurotransmitters (i.e., chemical messengers), but they do not induce changes in other tissues, as does induction.

## 23. D is correct.

*Acrosomal reaction* by the sperm is hydrolytic enzymes degrading the plasma membrane.

*Acrosome* is at the tip of sperm and contains specialized secretory molecules.

*Acrosomal reaction* is due to a signaling cascade involving the glycoproteins on the egg's surface.

*Acrosin* digests the zona pellucida and membrane of the oocyte, and the sperm releases its degradation enzymes to penetrate the egg's tough coating and allow the sperm to bind and fuse with the egg.

## 24. C is correct.

*Endoderm* is the innermost germ layer and gives rise to the inner lining of the respiratory and digestive tracts and associated organs.

*Blastula* refers to a hollow ball of embryonic cells arising from the morula. The blastula is *not* a germ layer.

*Ectoderm* is the outermost germ layer and gives rise to the hair, nails, eyes, skin, and central nervous system.

## 25. E is correct.

*Inner cell mass* is the cells inside the primordial embryo that forms before implantation and gives rise to the definitive structures of the fetus.

*Primitive streak* is a structure that forms in the blastula during the initial stages of embryonic development.

Primitive streak establishes:

> *bilateral symmetry*
>
> determines the *site of gastrulation*
>
> *initiates germ layer formation*

　　　　　

**26. D is correct.**

*Fetal circulatory system* changes as newborns begin using their lungs at birth.

Resistance in pulmonary blood vessels decreases, causing an increase in blood flow to lungs.

*Umbilical blood flow* ceases at birth, and blood pressure in the inferior vena cava decreases, which causes a decrease in pressure in the right atrium.

In contrast, the left atrial pressure increases due to increased blood flow from the lungs.

Increased left atrial pressure, coupled with decreased right atrial pressure, causes closure of the foramen ovale.

> *Ductus arteriosus* (i.e., connects the pulmonary artery to the aorta) constricts and subsequently is sealed.

> *Ductus venosus*, which shunts blood from the left umbilical vein directly to the inferior vena cava to allow oxygenated blood from the placenta to bypass the liver, completely closes within three months after birth.

Fetus produces adult hemoglobin a few weeks before birth (2 α and 2 β chains; alpha and beta) though lower amounts of fetal hemoglobin continue until the production stops.

After the first year, low fetal hemoglobin levels (2 α and 2 γ chains; alpha and gamma) are in the infant's blood.

**27. B is correct.**

*Homeotic genes* encode the related homeodomain proteins involved in developmental patterns and sequences.

*Homeobox* is a stretch of DNA about 180 nucleotides long that encodes a homeodomain (i.e., protein) in vertebrates and invertebrates.

*Exons* (expressed sequences) are retained during RNA processing of the primary transcript (hnRNA) into mRNA for translation into proteins.

**28. A is correct.**

*Somatic cells* have the same genome, but individual cells express different genes.

Differences in expression are *spatial* (i.e., cell type) or *temporal* (i.e., stages of development).

*Ectoderm tissue* arises after gastrulation as a primary germ layer.

Each germ layer retains its *determination*.

*Somatic cells* have identical genomes compared to their parents. *Gametes* (via segregation and recombination during prophase I) have unique 1N genomes compared to their parents.

**29. C is correct.**

*Yolk sac i*n humans gives rise to *blood cells* and *gamete-forming cells.*

*Chorion* is a double-layered membrane of trophoblast and extra-embryonic mesoderm.

Chorion gives rise to the fetal part of the placenta.

*Luteal placental shift* (7-9 weeks) is when the placenta develops enough to produce hormones to sustain the pregnancy.

Before this shift, the *placenta secretes progesterone* instead of the *corpus luteum.*

In humans, waste goes to the placenta and is received by the mother.

Non-placental organisms use allantois to collect waste.

**30. A is correct.**

*Trophoblast* is primarily responsible for forming placental tissue.

Trophoblast gives rise to the outer layer of a blastocyst, provides nutrients to the embryo, and develops into a large part of the placenta.

Trophoblasts form during the first stage of pregnancy and are the first cells to *differentiate* from the fertilized egg.

**31. C is correct.**

*Spina bifida* is a bone abnormality that arises from the embryonic mesoderm germ layer.

A lesion to the mesoderm would affect the development of other structures based on different connective tissue (e.g., blood, blood vessels, muscles, and connective tissue of organs).

Thus, this lesion affects the development of blood vessels and muscles.

I: *intestinal epithelium* develops from *endoderm.*

II: skin, hair, and the nervous system develop from *ectoderm.*

**32. B is correct.**

Homeotic genes encode the related homeodomain proteins involved in *developmental patterns and sequences.*

*Homeotic genes* are involved in developmental patterns and sequences.

For example, homeotic genes determine where, when, and how body segments develop.

*continued...*

---

     

Alterations in these genes cause changes in body parts and structure patterns, sometimes resulting in dramatic effects.

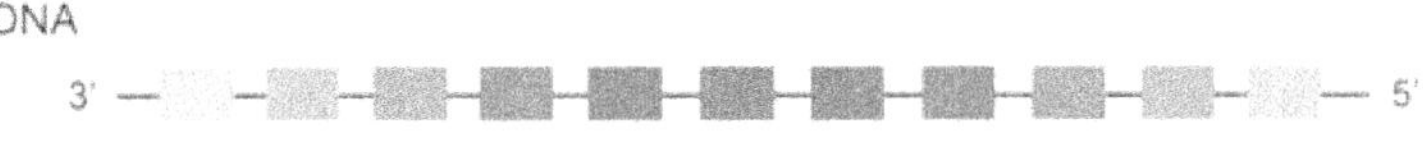


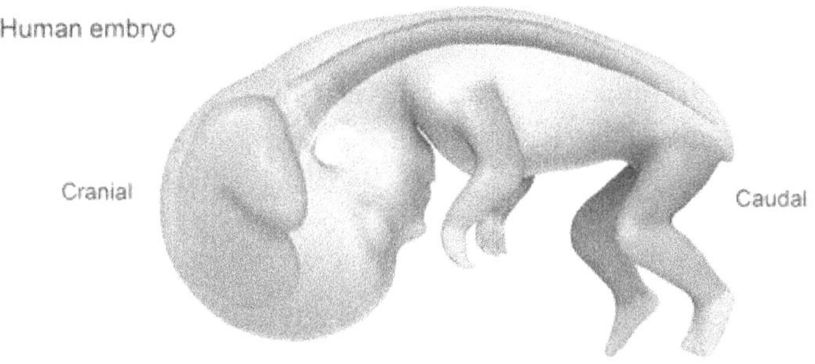


*HOX genes encode homeodomain proteins as developmental regulators of the anterior-posterior axis. The linear sequence of genes on the chromosome corresponds to the cranial-caudal orientation*

*Homeodomain* is a protein structural domain that binds DNA or RNA and is a transcription factor (i.e., facilitates polymerase binding and activation).

## 33. D is correct.

*Mesoderm* forms muscles, blood, bone, reproductive organs, and kidneys.

## 34. C is correct.

*Primitive streak* is a transitional structure formed at the onset of gastrulation.

*Inner cell mass* is converted into the trilaminar embryonic disc comprised of the three germ layers: ectoderm, mesoderm, and endoderm.

*Primitive streak formation* is an early process of embryonic development.

## 35. A is correct.

During labor, the *oxytocin* hormone stimulates contractions of uterine smooth muscle.

## 36.  D is correct.

*Ectoderm cells* are determined but not differentiated because they have the potential to develop into more than one type of tissue, but not any type.

Experimentally, ectoderm development was influenced (i.e., induced) by changing locations within the developing embryo. The underlying *mesoderm* differentiated because it released the molecular inducers as signals to the overlying ectoderm (undifferentiated).

*Ectoderm* cell *location* is essential because their development (i.e., differentiation into specific tissue/structures) is induced by the underlying mesoderm cells that send chemical substances (i.e., inducers) specific to the position of the mesoderm.

Mesoderm, not ectoderm, determines cellular differentiation at this stage of development.

*Differentiated fate* of the cells has not yet been determined because the transplanted cells would develop into wing feathers instead of claws.

Ectoderm cells cannot develop into any tissue because location influences the development of transplanted ectoderm cells.

## 37.  D is correct.

Recognition of the sperm by the *vitelline envelope* (i.e., a membrane between the outer zona pellucid and inner plasma membrane) triggers the cortical reaction, transforming it into the hard *fertilization membrane* as a physical barrier to other spermatozoa (i.e., *slow block to polyspermy*).

*Cortical reaction* within the egg is analogous to the acrosomal reaction within the sperm.

## 38.  C is correct.

*Polyspermy* in humans results in a nonviable zygote.

## 39.  E is correct.

*Mesoderm germ layer* gives rise to many tissues.

*Intestinal mucosa* is derived from *endoderm*

*Nerve* is derived from *ectoderm*.

*Lung epithelium* is derived from *endoderm*.

**40. C is correct.**

*Genomic imprinting* is *epigenetic* (i.e., *heritable change*) when specific genes are expressed in a parent-of-origin-specific manner.

Genomic imprinting is an *epigenetic process* involving *DNA methylation* and *histone modulation* to achieve single-allelic gene expression without altering the original genetic sequence.

**41. C is correct.**

Embryos require *greater protein translation rates* using ribosomes to read mRNA transcripts.

**42. E is correct.**

*Zygote* (i.e., fertilized egg) is the first structure to form during human fertilization that undergoes rapid cell divisions without significant cellular growth to produce a cluster of cells of the same size as the original zygote.

Cells derived from cleavage are *blastomeres* and form a solid mass known as a *morula.*

*Cleavage ends* with the formation of the *blastula.*

> Ovulation → fertilization (sperm and oocyte) → diploid zygote
> (undergoes cleavage) → morula (solid ball of cells) → blastocoel
> (undergoes invagination) → blastopore (ectoderm and endoderm)

**43. A is correct.**

*Capacitation* is the final step in spermatozoa maturation. It is required for competence to fertilize oocytes.

Capacitation destabilizes the acrosomal sperm head membrane, allowing greater binding between sperm and oocyte by removing steroids (e.g., cholesterol) and non-covalently bound glycoproteins, which increases membrane fluidity and $Ca^{2+}$ permeability.

$Ca^{2+}$ influx produces intracellular cAMP levels and increased sperm motility.

**44. C is correct.**

*Endoderm* develops into *epithelial linings* of the digestive and respiratory tracts, parts of the liver, thyroid, pancreas, and bladder lining.

**45. E is correct.**

*Homeotic genes* influence the development of specific structures in plants and animals, such as the Hox and ParaHox genes, which are essential for segmentation.

Homeotic genes determine where, when, and how body segments develop in organisms.

Alterations in these genes cause changes in patterns of body parts, sometimes causing dramatic effects, such as legs growing in place of antennae or an extra set of wings.

*Loss-of-function* mutations result in the gene product with less or no function.

An allele with loss of function (i.e., null allele) is an *amorphic* (i.e., complete loss of gene function) mutation.

Phenotypes associated with such mutations are often recessive.

Exceptions are haploid or *haploinsufficiency* (i.e., a diploid organism has a single copy of the functional gene) when the reduced dosage of a normal gene product is insufficient for a typical phenotype.

**46. A is correct.**

Cells may assume several fates and are not yet terminally *differentiated.*

*Gastrula cells* influenced by their surroundings are *competent.*

*Ectoderm layer* gives rise to the eye, among other structures.

Cells can become other ectoderm tissue (e.g., gills).

**47. B is correct.**

*Proteases* and *acrosin* enzymes degrade the protective barriers around the egg and allow the sperm to penetrate.

*Acrosomal reaction* by the sperm is hydrolytic enzymes degrading the plasma membrane.

*Acrosome*s are at the tip of the sperm and contain specialized secretory molecules.

Acrosomal reaction is a signaling cascade involving the glycoproteins on the egg's surface.

*Acrosin* digests the zona pellucida and membrane of the oocyte.

Sperm releases its degradation enzymes to penetrate the egg's tough coating and allow the sperm to bind and fuse with the egg.

**48. A is correct.**

*Human blastocyst implants* in the uterine wall about a *week* after fertilization.

## 49.  B is correct.

*Ectoderm* develops into the nervous system, the epidermis, the eye lens, and the inner ear.

*Endoderm* develops into the lining of the digestive tract, lungs, liver, and pancreas.

*Mesoderm* develops into the connective tissue, muscles, skeleton, circulatory system, gonads, and kidneys.

## 50.  A is correct.

After the infant's first breath, the newborn's cardiovascular system constricts the *ductus arteriosus* (i.e., connects pulmonary artery to aorta) and converts it to *ligamentum arteriosum*.

*Fetal circulatory systems* change at birth as the newborn uses its lungs.

Resistance in pulmonary blood vessels decreases, increasing blood flow to the lungs.

At birth, umbilical blood flow ceases, and blood pressure in the inferior vena cava decreases, which causes a decrease in pressure in the right atrium.

In contrast, the left atrial pressure increases due to increased blood flow from the lungs.

Increased left atrial and decreased right atrial pressure causes *closure* of the *foramen ovale*.

*Ductus venosus*, which shunts blood from the left umbilical vein directly to the inferior vena cava to allow oxygenated blood from the placenta to bypass the liver, completely closes within three months after birth.

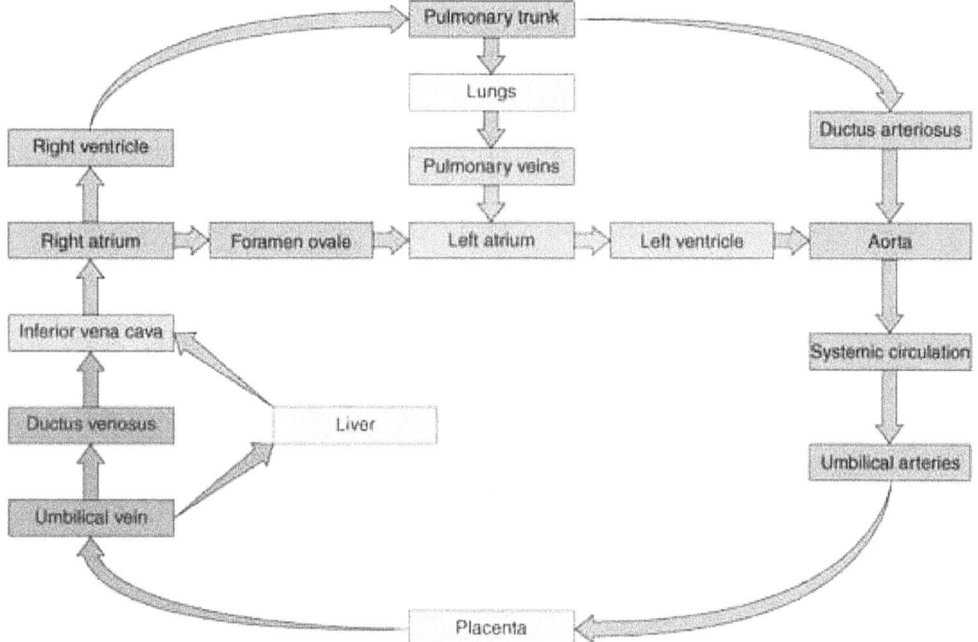


*Adult and fetal circulatory systems: ductus arteriosus and ductus venosus provide shunt pathways*

**51.  C is correct.**

*Genomic imprinting* is an *epigenetic* (i.e., heritable changes in gene activity) process involving DNA methylation and histone remodeling for monoallelic (i.e., single allele) gene expression without altering the genetic sequence within the genome.

*Epigenetic changes* are established in the germline (i.e., cells that give rise to gametes of egg/sperm) and can be maintained through mitotic divisions.

**52.  E is correct.**

*Amniotic fluid*, surrounded by the amnion, is a liquid environment around the egg that protects it from shock.

*Embryonic membranes* include:

> 1) **chorion** lines the inside of the shell and permits gas exchange;

> 2) **allantois** is a saclike structure developed from the digestive tract and functions in respiration, excretion, and gas exchange with the external environment;

> 3) **amnion** encloses *amniotic fluid* as a watery environment for embryogenesis and shock protection;

> 4) **yolk sac** encloses the yolk and transfers food to the developing embryo.

**53.  A is correct.**

*Mesoderm* gives rise to the *entire circulatory system, muscle,* and most other tissue between the *gut and skin* (excluding the nervous system).

*Ectoderm* gives rise to skin, the nervous system, the retina, and the lens.

*Differentiated endoderm* has a restricted fate that does not include heart tissue.

*Endoderm* gives rise to the *inner lining of the gut.*

**54.  C is correct.**

At the end of the *first trimester*, an ultrasound can determine the sex of the fetus.

**55.  E is correct.**

*Eggs are viable* and can be fertilized 12-24 hours after ovulation.

  

## 56.  B is correct.

*Blastulation* begins when the *morula* develops a fluid-filled cavity (i.e., blastocoel).

By the fourth day, it becomes a *hollow sphere of cells* (i.e., blastula).

*Gastrula* is the embryonic stage characterized by the three primary germ layers (i.e., endoderm, ectoderm, and mesoderm), the blastocoel, and the archenteron.

Early gastrula is two-layered (i.e., ectoderm and endoderm); a third layer (i.e., mesoderm) develops shortly afterward.

Gastrulation is followed by *organogenesis,* when individual organs develop within newly formed germ layers.

*Morula* is the solid ball of cells from the early cleavage stages in the zygote.

*Zygote* is the (2N) cell formed by the fusion of two (1N) gametes (e.g., ovum and sperm).

## 57.  B is correct.

*Proteases* and *acrosin* enzymes degrade the protective barriers around the egg and allow the sperm to penetrate.

*Acrosomal reaction* by the sperm is hydrolytic enzymes degrading the plasma membrane.

*Acrosome*s are at the tip of the sperm and contain specialized secretory molecules.

*Acrosomal reaction* is due to a signaling cascade involving the glycoproteins on the egg's surface.

*Acrosin* digests the zona pellucida and membrane of the oocyte.

Sperm releases its degradation enzymes to penetrate the egg's tough coating and allow the sperm to bind and fuse with the egg.

## 58.  C is correct.

Regarding fertilization, the *vagina's acidic environment* destroys millions of sperm cells.

**59. B is correct.**

*Zygote* is a fertilized egg undergoing rapid cell divisions without significant cellular growth to produce a cluster of cells of the same total size as the original zygote.

Cells derived from cleavage are *blastomeres* and form a solid mass known as a *morula*.

*Cleavage ends* with the formation of the *blastula*.

Among species, depending mainly on the amount of yolk in the egg, cleavage is:

> *holoblastic* (i.e., total or entire cleavage) or

> *meroblastic* (i.e., partial cleavage).

Egg pole with the highest yolk concentration is the *vegetal pole*, while the opposite is the *animal pole*.

Humans undergo *holoblastic cleavage* (i.e., total cleavage).

**60. B is correct.**

*Fertilized egg* is a *zygote* undergoing rapid cell division without significant growth, producing a cluster of cells the same total size as the original zygote.

Cells derived from zygote cleavage are *blastomeres* and form a solid mass as a *morula*.

Cleavage ends with the formation of the *blastula* (i.e., hollow ball of cells).

In varied species, mainly depending on the amount of yolk in the egg, cleavage is *holoblastic* (i.e., total or entire cleavage) or *meroblastic* (i.e., partial cleavage).

> *Vegetal pole* is the egg pole with the *highest* yolk concentration.

> *Animal pole* is the egg pole with the *least* yolk concentration.

Humans undergo *holoblastic cleavage*.

*Notes for active learning*

# Frank J. Addivinola, Ph.D.

The lead author and chief editor of this study guide is Dr. Frank Addivinola. With his outstanding education, laboratory research, and decades of university science teaching, Dr. Addivinola lent his expertise to develop this book.

Dr. Frank Addivinola conducted original research in developmental biology as a doctoral candidate and pre-IRTA fellow in Molecular and Cell Biology at the National Institutes of Health (NIH). His dissertation advisor was Nobel laureate Marshall W. Nirenberg, Chief of the Biochemical Genetics Laboratory at the National Heart, Lung, and Blood Institute (NHLBI). Before NIH, Dr. Addivinola researched prostate cancer in the Cell Growth and Regulation Laboratory of Dr. Arthur Pardee at the Dana Farber Cancer Institute of Harvard Medical School.

Dr. Addivinola holds an undergraduate degree in biology from Williams College. He completed his Masters at Harvard University, Masters in Biotechnology at Johns Hopkins University, and five other graduate degrees at the University of Maryland University College, Suffolk University, and Northeastern University.

During his extensive career, Dr. Addivinola held faculty positions at colleges and universities, including Harvard University, Johns Hopkins University, University of Maryland and Northeastern University, and taught numerous undergraduate and graduate-level courses in biology, biochemistry, organic chemistry, inorganic chemistry, anatomy and physiology, medical terminology, nutrition, and medical ethics. He received several awards for his research and presentations.

## *Essential Biology Self-Teaching Guides*

Eukaryotic Cell & Cellular Metabolism

Molecular Biology & Genetics

Nervous & Endocrine Systems

Circulatory, Respiratory & Immune Systems

Digestive & Excretory Systems

Muscle, Skeletal & Integumentary Systems

Reproduction & Development

Microbiology

Plants & Photosynthesis

Evolution, Classification & Diversity

Ecology & Population Biology

**Visit our Amazon store**

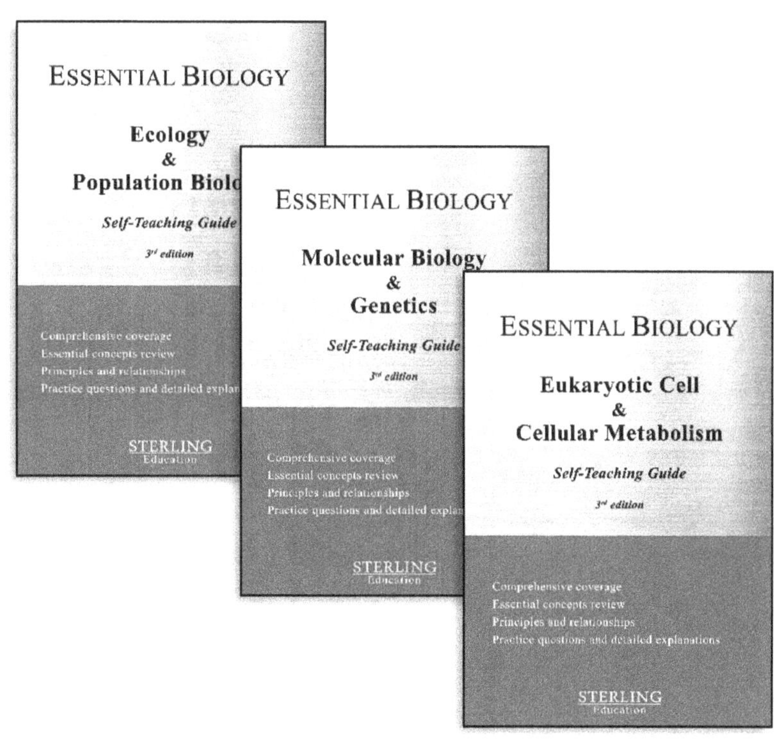